AF881339

FM 101-5-2

Field Manual
No. 101-5-2

Headquarters
Department of the Army
Washington, DC, 29 June 1999

US ARMY REPORT AND MESSAGE FORMATS

Contents

DISTRIBUTION RESTRICTION: Approved for public release; distribution is unlimited.

PREFACE

FM 101-5-2 is the Army's capstone manual for standardized report and message formats. It provides a standard, readily available reference from which soldiers as well as automation designers can extract report and message templates. FM 101-5-2 is a compendium of formats commonly used by tactical units from small unit to corps and forms the baseline for reporting and communicating as command, control, and communications technology evolves. It allows a common, authoritative understanding of reporting and communicating to exist among all US Army elements. It also relates to, supports, and drives command, control, and communications doctrine and US Army interaction with the joint and multinational communications communities.

FM 101-5-2 facilitates the tailoring and task organization of US Army elements through standardization of the command, control, and communications formats units use to communicate internally. The voice-message templates contained in FM 101-5-2 provide the bridge between technologically advanced units and those not yet modernized. This is significant because, while our modernized units are among the most capable, nonmodernized units currently comprise the majority of the US Army's active component, all of the reserve components, and the majority of our potential allies. The US Army will continue to modernize all units, but the requirement to communicate expeditiously and succinctly via voice will remain. Potential attacks on our command, control, and communications networks, equipment damage, incompatible communications systems, and equipment failures are other consequences that may require the use of voice message formats, even for modernized units.

As the US Army's doctrinal library for report and message voice templates, FM 101-5-2 is intended to prevent units from wasting time and resources designing command and control formats. Units at different locations and echelons can develop similar SOPs for report and message formats (voice and digital) that facilitate command, control, communications, computer, and intelligence (C4I) exchanges. The FM 101-5-2 formats are derived from existing Army Battle Command System (ABCS), United States Message Text Formats (USMTF), Allied Procedures Publication- 9 (APP-9) formats, as well as numerous unit SOPs. These sources, especially the formats from existing unit SOPs, have been collected, discussed, and chosen as the most user-friendly, functional, and adaptable.

As a user's manual, FM 101-5-2 will drive the formatting of future ABCS report and messages, US Army input to the development of APP-9, and the USMTF message formats. Formats contained in FM 101-5-2 are the mechanism through which the US Army seeks to adapt these joint and multinational formats into a user-friendly and standardized land component message format library. Parallel formats for ABCS, APP-9 and USMTF will enable US Army units to operate in joint and multinational command structures without reconfiguring their normal mode of operation. All future digitization formats for the US Army will be based on these voice message templates. FM 101-5-2 supports the efforts of FM 100-34-1 (Tactics, Techniques, and Procedures for Command Post Operations- TBP) and is consistent with current and emerging US Army and joint doctrine.

FM 101-5-2 is directive and prescriptive. Units will use its report and message formats for all tactical communications. Units will not modify these formats unless authorized by the unit commander for critical information requirements. For Army-wide use, this manual does not include many branch-specific or technical reports and messages. These formats are contained in field manuals published by their respective proponents. Branch-specific or technical reports complement the reports in this manual and are not considered superseded by it. However, use the formats in FM 101-5-2 whenever possible.

Headquarters, Combined Arms Center is the proponent for this publication. Send comments and recommended changes on DA Form 2028 to the Commander, US Army Combined Arms Center and Fort Leavenworth, ATTN: ATZL-SWW-D, Fort Leavenworth, KS 66027-6900.

Unless otherwise stated, masculine nouns or pronouns do not refer exclusively to men; however, reports refer to messages.

SECTION I—INSTRUCTIONS

The report and message formats in this manual are designed to ease preparation and manual transmission of written and voice reports and messages. They provide an organized template to manually record, pass, and store information. To send a formatted voice message, operators first fill in the appropriate blanks on the required message template and then read the message over the radio or telephone. The receiver, knowing the message type he is receiving, can easily record the information in the proper format. Voice messages normally are used when record traffic devices are inoperative or not available, or when the tactical situation does not allow adequate time to transmit a record message.

FM 101-5-2 enables standardization of battlefield reporting across all levels of command and throughout different types of units. Every format listed in FM 101-5-2 operates as a voice message format. Using standardized voice formats permits all units to communicate regardless of the unit's level of modernization. It is anticipated that future digital formats will evolve from those found in FM 101-5-2.

FM 101-5-2 has three sections: Section One contains instructions on transmitting written and voice reports and messages. Section Two contains message formats. Its use is discussed below. Section Three contains cross-over indexes. These indexes list the names, numbers, and locations in this manual of reports commonly associated with selected subject areas.

Message Organization

All radio-telephone report and message formats consist of three main parts: heading, body, and conclusion. Figure I-1 shows an example of the parts of a message. FM 24-19, Chapter 5, discusses message formatting in detail.

The *heading* is the administrative portion of the format. Its purpose is to identify the message source, destination, type, and importance. The heading contains the message addressee, originator, and precedence. It also includes the message classification, if required. (All report and message formats in FM 101-5-2 are unclassified. The unit commander or SOP determines the classification of a completed message.) The completed heading is normally included when sending a hard-copy message, as when sending by TACFAX, courier, or MCS. However, it must also be included in voice messages. Figure I-2 gives an example of the standard US Army message heading and explains how to transmit it by voice.

The *body* contains the information the originator wants to send to the addressee. In a properly prepared message the information is in the line-by-line format established for the message type. This information exchange is the main focus of FM 101-5-2. Figure I-3 shows an example of a message body.

The *conclusion* consists of the message authentication. An authentication is normally an alphanumeric from the unit signal operating instructions (SOI). The last line of every format in this manual is the authentication line (the conclusion) of the message. Upon receiving a complete message that is properly authenticated, the addressee indicates this with a response similar to the one in Figure I-4. If the originator requests an acknowledgement, the addressee replies by transmitting the proword "*WILCO,*" Lines 1 and 2 of the message being acknowledged, and the proword "*OUT.*" If the originator requires an acknowledgement and read-back, the addressee replies by transmitting "*WILCO*" and repeating the lines for which the read-back is required followed by "*OUT.*"

How to Use This Manual

Section Two contains formats listed alphabetically by message name. The body of each format is preceded by the report's short title (for example, SIR for Serious Incident Report), report number, and a paragraph of general instructions.

Report Formats Numbering System. Each format has a report number. Report numbers beginning with the letters A through M indicate that the origin of this message is from APP-9 or DA PAM 25-7. Report numbers beginning with Z indicate that the report was taken from a unit SOP or produced by the Combined Arms Doctrine Directorate of the US Army Command and General Staff College at Fort Leavenworth, Kansas. To maintain standardization across the US Army, do not change format numbers.

General Instructions. Instructions briefly describe the format's use and list a reference that provides more detail. When a message is sent at a precedence higher than ROUTINE, that is noted. Message precedence is indicated in all upper case letters.

HEADING

UNCLASSIFIED
FM 101-5-2 VOICE MESSAGE FORMATS
IMPLDATE: Pending

VOICE MESSAGE TEMPLATE MESSAGE NUMBER: F541
TITLE: SPOT REPORT [SPTRPT]

Initially the originator makes radio contact with the addressee.

"Alpha 09" THIS IS "India 86" " Over"

"INDIA 86" THIS "Alpha 09" "Over"

Once communication is established Originator responds:

"*Alpha 09*" (addressee) THIS IS "*India 86*" (originator) "*SPOT REPORT/ OVER*"

This allows the addressee time to be ready to copy transmitted report or message.

"India 86" (originator) THIS IS **"Alpha 09"** (addressee) **"ROGER SEND IT"/ OVER"**

At this time the originator will send the report or message IAW with the FM 101-5-2 Formats:

"Alpha 09" (addressee) THIS IS *"India 86"* (originator)

(Circle and transmit option selected IAW unit SOP)

FLASH IMMEDIATE PRIORITY ROUTINE (Underline and transmit the precedence of this message.)

BODY

TOP SECRET SECRET CONFIDENTIAL (Underline and transmit the security classification of this message.)

"ROUTINE/ (or slash spoken) SECRET"

"LINE 1 – TIME "091500SEP 97" (Date and Time Group of Report)

"LINE 2 – UNIT "B CO 3-66AR" (Unit Making Report)

"LINE 3 – SIZE "Two groups of three" (Of observed activity)

"LINE 4 – ACTIVITY "Moving north to south in wedge formation (Observed unit activity)

"LINE 5 – LOCATION "A(alpha)R(Romeo)121989" (Location of Unit Observed, if Necessary Reporting Unit Location)

"LINE 6 – UNIT "Two Enemy Tank Platoons unknown" (Enemy Unit)

"LINE 7 – TIME "091445 SEP 97" (Time of Observation)

"LINE 8 – EQUIPMENT "Enemy tank platoon had snorkel device mounted" (Equipment of Unit Observed)

"LINE 9 – SENDERS ASSESSMENT "Enemy CRP attempting to locate a ford site across cowhouse creek vic CP 09" (Why)

"LINE 10 – NARRATIVE "Have begun to initiate fire support" (Free Text for Additional Information Required for Clarification of Report)

CONCLUSION

"LINE 11 – AUTHENTICATION "JWA" (Report Authentication)
"OVER"

"India 86 THIS Alpha 09 Acknowledged OUT"

Figure I-1. Voice Message Template

Line Number And Name. Each format lists the information the report is used to transmit by line number, line name, and a description of the type of information.

- When sending voice messages, the line number is normally transmitted. The line name is not transmitted unless the name itself is necessary information. For example, Line 3 of an Acknowledge message (AKNLDG) is the message type and is therefore transmitted. Lines 2 and 3 of an Acknowledge message would sound like this when sent by voice (over a secure net): "Line 2, 2-39 FA; Line 3, Acknowledge."

- When standard information is associated with a line, it is placed in the right column of the format in upper-case letters. For example, the standard entries for line 6 of an Acknowledge message are ACKNOWLEDGE and WILCO.

- When a line requires detailed instructions, two asterisks precede the instructions (**).

- When transmitting information on one line makes the following line redundant, the word "(OR)" is placed between the lines. Do not transmit redundant lines.

- Each format contains lines that require some of the information discussed below. Follow these instructions when preparing messages requiring transmission of these types of information.

Date and Time Group (DTG). The DTG line of a message identifies the message origination time. Additionally, many report and message formats require DTG for additional information. Regardless of where the DTG is used within a report or message, it is expressed as DDHHMM(X) MMMYY where-

- DD indicates the day of the month.

- HH indicates the hour of the day, using the 24-hour clock.

- MM indicates the minutes within the hour.

- X indicates the alphabetic character that denotes the time zone. In written messages, the time zone is followed by a space. The time zone indicator will be local time zone unless specified in the unit's operation order or in the FM 101-5-2 format. As an example, DD 1972, *Joint Attack Air Request*, requires that times be sent as Greenwich Mean, or "zulu" time, which is indicated with the letter Z.

- MMM indicates the first three letters of the month of the year. In written messages, all letters are in upper case.

- YY indicates the last two digits of the year.

- For example, 3 PM local time on 9 September 1997 would be expressed as "091500(L) SEP 97." In a voice message it would be said: "Zero niner one five hundred lima (or local) September niner-seven."

Units. The first line of a format requiring an entry for "unit" is to identify the unit making the report. Occasionally, a subunit line is also provided for reporting information from or about subunits. Unit SOPs dictate the format for the unit name; for example, call sign.

Locations. Use the Military Grid Reference System (MGRS) to express all locations unless the format specifies a different coordinate system such as latitude and longitude (LAT/LONG) or Universal Transverse Mercator (UTM). A complete MGRS location includes a longitudinal zone designated by a number (for example, 32), a latitudinal band designated by a letter (for example, T), a 100,000 meter grid square designated by two letters (for example AR), and a grid coordinate designated by digits. The **UTM Grid Zone Designator** consists of the longitudinal zone and latitudinal band (for example, 32T). The **MGRS Grid Zone Designator** is the two-letter 100,000 meter designator (for example, AR). (See FM 21-26 for further details on MGRS.)

Send all locations as six-digit grids preceded by the 100,000 meter grid square designator, for example "AR121989" unless instructions state otherwise. To avoid confusion do not transmit the zone and band indicators unless requested or required in the message format. If requestors require locations expressed to a greater precession than a six-digit coordinate, they must specify the required detail prior to the information exchange.

Addition or Deletion of Line Numbers. Originators of voice messages complete only those lines required to pass the essential information, omitting lines that contain no information unless a negative report is necessary. Units may add lines or repeat lines to complete their information requirements. While actual message transmissions may vary slightly from the format in this manual, do not delete lines from formats when reproducing them in unit SOPs to maintain standardization across the US Army.

UNCLASSIFIED
FM 101-5-2 VOICE MESSAGE FORMATS
IMPLDATE: Pending

VOICE MESSAGE TEMPLATE MESSAGE NUMBER: F541
TITLE: SPOT REPORT [SPTRPT]

Initially the originator makes radio contact with the addressee. Italicized words demonstrate how to speak a template over the radio.

"Alpha 09 THIS IS India 86 Over"

"India 86 THIS IS Alpha 09 Over"

Once communication is established, the Originator responds:

"Alpha 09 ____________ THIS IS India 86 ____________________ SPOT REPORT... OVER"
addressee originator

This allows the addressee time to copy a transmitted report or message.

"India 86 THIS IS Alpha 09 ROGER SEND IT...OVER"
originator addressee

"Alpha 09 ____________ THIS IS India 86"
addressee originator

(Circle and transmit option selected IAW unit SOP)

FLASH IMMEDIATE PRIORITY ROUTINE
(Underline and transmit the precedence of this message.)

TOP SECRET SECRET CONFIDENTIAL
(Underline and transmit the security classification of this message.)
"*ROUTINE/SECRET*" (the slash is spoken)

Figure I-2. Example of Message Heading

"*LINE 1* – DATE AND TIME: *091500SEP 97*" (DTG)

"*LINE 2* – UNIT: *B CO 3-66 AR*" (Unit Making Report)

"*LINE 3* – SIZE: *Two groups of three*" (Of Observed Activity)

"*LINE 4* – ACTIVITY: *Moving north to south in wedge formation*" (Observed Unit Activity)

"*LINE 5* – LOCATION: *A(alpha)R(Romeo)121989*" (MGRS or Six-Digit Grid Coordinate With Grid Zone Designator)

"*LINE 6* – UNIT: *Two Enemy Tank Platoons unknown*" (Enemy Unit)

"*LINE 7* – TIME : *091445 SEP 97*" (Time of Observation)

"*LINE 8* – EQUIPMENT: *Enemy tank platoon had snorkel device mounted*" (Equipment of Unit Observed)

"*LINE 9* – SENDERS ASSESSMENT: *Enemy CRP attempting to locate ford site across Cowhouse Creek vic CP 09*" (Why)

"*LINE 10* – NARRATIVE: *Have begun engaging with indirect fires*" (Free Text for Additional Information Required for Clarification of Report)

"*LINE 11* – AUTHENTICATION: *JWA OVER*" (Report Authentication)

Figure I-3. Sample Message Body

"*India 86 THIS IS Alpha 09. Acknowledged. OUT*"

Figure I-4. Sample Message Conclusion

SECTION II—VOICE MESSAGE TEMPLATE FORMATS

TITLE: ACCIDENT REPORT/SERIOUS INCIDENT REPORT [SIR]
REPORT NUMBER: A001

GENERAL INSTRUCTIONS: Use to convey flash traffic to the commander and command post relating to an accident or serious incident within the command. Use FLASH precedence on CMD nets. For all other stipulations and instructions refer to AR 190-40 (SIR) and AR 385-40 (Accident Report).

LINE 1 – DATE AND TIME ________________________________ (DTG of Report)

LINE 2 – UNIT ________________________________ (Unit Making Report)

LINE 3 – CATEGORY ________________________________ (Category 1 or 2)

LINE 4 – TYPE ________________________________ (Type of Incident)

LINE 5 – TIME OF INCIDENT ________________________________ (DTG of Incident)
WEEK END/HOLIDAY ________________________________ (Indicate YES or NO and Include Holiday Name)

LINE 6 – LOCATION ________________________________ (UTM or Six-Digit Grid Coordinate With MGRS Grid Zone Designator)

LINE 7 – PERSONNEL ________________________________ (Personnel Involved)

a. (SUBJECT)
(1)________________________(Name, Injury, Fatality)
(2)________________________(Pay Grade)
(3)________________________(SSN)
(4)________________________(Race)
(5)________________________(Sex)
(6)________________________(Age)
(7)________________________(Position (CDR, PSG, Gunner))
(8)________________________(Security Clearance)
(9)________________________(Unit, Station, and MACOM Assigned)
(10)________________________(Duty Status (Leave))

b. (VICTIM)
(1)________________________(Name, Injury, Fatality)
(2)________________________(Pay Grade)
(3)________________________(SSN)

**Continued on next page.

TITLE: ACCIDENT REPORT/SERIOUS INCIDENT REPORT [SIR] (CONTINUED)
REPORT NUMBER: A001

(4)________________________(Race)
(5)________________________(Sex)
(6)________________________(Age)
(7)________________________(Position (CDR, PSG, Gunner))
(8)________________________(Security Clearance)
(9)________________________(Unit, Station, and MACOM Assigned)
(10)_______________________(Duty Status (Leave))

**Repeat lines 7a (1-10) through 7b (1-10) up to four times to transmit multiple data sets. Assign sequential line numbers to succeeding iterations; for example, first iteration a (1) through a (10) or b (1) through b (10); second iteration a (11) through a (20) or b (11) through b (20); third iteration a (21) through a (30) or b (21) through b (30); fourth iteration a (31) through a (40) or b (31) through b (40).

LINE 8 – SUMMARY ________________________________ (Summary of Incident)

LINE 9 – PUBLICITY ______________________________ (Adverse or Anticipated Publicity)

LINE 10 – COMMANDER ___________________________ (Commander Reporting)

LINE 11 – POC______________________________________ (Unit POC/Duty Position)

LINE 12 – DOWNGRADE_____________________________ (Downgrading Instructions)

LINE 13 – NARRATIVE _____________________________ (Free Text for Additional Information Required for Clarification of Report

LINE 14 – AUTHENTICATION ________________________ (Report Authentication)

ACCIDENT SUPPLEMENT (GROUND (GRND)/AIRCRAFT (AC))

LINE 14A –POC_____________________________________ (E-mail, Phone, DSN)

LINE 14B –MACOM ________________________________ (TRADOC, FORSCOM, USAREUR)

LINE 14C – ACCIDENT CLASS_________________________ (GRND/AC—A or B AC—C)

LINE 14D –WEATHER ______________________________ (Visibility, Light Level, Precipitation, Wind)

**Continued on next page.

TITLE: ACCIDENT REPORT/SERIOUS INCIDENT REPORT [SIR] (CONTINUED)
REPORT NUMBER: A001

ACCIDENT SUPPLEMENT (GROUND (GRND)/AIRCRAFT (AC)) (CONTINUED)

LINE 14E – NIGHT VISION DEVICE ________________________	(Yes, Type, Nomenclature. For Example, NVG-AN/PVS-5, Abrams Thermal Viewer, No)
LINE 14F – MILITARY INSTALLATION ____________________	(Nearest Site)
LINE 14G – EXPLOSIVE, HAZARDOUS MATERIAL ___________	(Involved: Yes or No, Secured: Yes or No)
LINE 14H – ACCIDENT SITE SECURED IAW AR 385-40 ________	(Yes or No)
LINE 14I – ACCIDENT SITE DISTURBED___________________	(Yes or No)
LINE 14J – PHOTOS OF DISTURBING SCENE________________	(Yes or No)
LINE 14K – FLIGHT DATA RECORDER INSTALLED___________	(Yes or No)
LINE 14L – LOCAL ACCIDENT INVESTIGATION BOARD APPT _	(Yes or No)
LINE 14M – NEAREST AIRFIELD __________________________	(4000 Ft. Min. Runway)
LINE 14N – NEAREST COMMERCIAL AIRFIELD_____________	(Name, Town, State)
LINE 14O – TYPE OF EQUIPMENT _________________________	(M1 Abrams, AH64, 30KW, Gen)
LINE 14P – AC SERIAL NUMBER __________________________	(Serial, Tail Number)
LINE 14Q – TYPE MISSION________________________________	(Training, Svc, Single Ship, Multi Ship)
LINE 14R – NAP OF THE EARTH___________________________	(Yes or No)
LINE 14S – FLIGHT RULES_______________________________	(VFR/IFR)

WORKSHEET FOR
TELEPHONIC NOTIFICATION OF AVIATION ACCIDENT / INCIDENT

For use of this form, see AR 385-40: the proponent agency is OCSA

SHADED BLOCKS ARE FOR USASC USE ONLY	A. ASMIS CASE NUMBER	B. TIME & DATE OPERATIONS RECEIVED REPORT			
NOTE: ITEMS 24 AND 25 ARE NOT REQUIRED FOR CLASS C ACCIDENT		a. Year	b. Month	c. Day	d. Time (local)

1. POINT OF CONTACT FOR ACCIDENT INFORMATION — a. Name: John Jones

b. Duty ☐ Commander ☐ Safety Officer ☐ Other *(Specify)* — c. Phone Number DSN: Commercial: 913-555-2569

2. ACCIDENT CLASSIFICATION	3. TIME & DATE OF ACCIDENT				4. AIRCRAFT SERIAL NUMBER	5. TYPE OF AIRCRAFT
☐ A ☐ B ☒ C	a. Year 98	b. Month Jun	c. Day 15	d. Time (local)		UH 1

6. PERIOD OF DAY	7. MISSION BEING PERFORMED		8. NOE
☐ Dawn ☐ Day ☐ Day ☐ Night	a. Type (Training, Svc, etc.) Training	b. Operation ☐ Single-Ship ☐ Multi-Ship	☐ Yes ☐ No

9. NIGHT VISION DEVICE		10. UNIT OWNING AIRCRAFT	11. MACOM
a. In Use ☐ Yes ☐ No	b. If Yes ☐ ANVIS ☐ AN/PVS-5 ☐ FLIR ☐ LLTV	4-44 AV Bn, 21 AR Div	IV Corps

12. MILITARY INSTALLATION NEAREST ACCIDENT SITE	13. EXACT ACCIDENT LOCATION
Camp Sherman, MO	VS15962847

CHECK "YES" or "NO" FOR QUESTIONS 13 THROUGH 17	Yes	No
14. EXPLOSIVE/HAZARDOUS/SENSITIVE MATERIALS INVOLVED?		X
15. IF YES TO #13, ARE THEY SECURE?		
16. ACCIDENT SITE SECURED IAW DA PAM385-40?	X	
17. HAS ACCIDENT SITE BEEN DISTURBED?		X
18. IF YES TO #16, WERE PHOTOS, ETC. MADE BEFORE DISTURBING THE SCENE?		
19. FLIGHT DATA RECORDER INSTALLED?		X
20. CLEARANCE WAS: ☐ VFR ☐ IFR		

21. PERSONNEL INVOLVED — a. No. of Personnel by Rank/Category: 1 Officer 1 WO; 1 Enlisted ____ Army Civilian; ____ Non-Army Civilian. b. Total No. of Personnel: 3

22. INJURIES (Enter # of each): ____ Fatalities; ____ Non-Fatal Injuries

As soon as possible, the following Additional information is required On all injured personnel; name, Personnel classification, degree of Injury and SSAN.

23. ACCIDENT SYNOPSIS (What Happened)

While on a training mission, the flight developed mechanical problems and made an emergency landing in the field located SW of the intersection of US Hwy 88 and US Hwy 11. It was a hard landing and damage to the aircraft is expected to exceed $10,000. No one was injured. Specific cause of mechanical failure has not been determined.

24. NEWS MEDIA AWARE OF ACCIDENT	25. NEAREST AIRFIELD
☐ Yes ☐ No	a. Nearest that can handle C-12 (4 ,000 ft. min.) b. Nearest commercial airfield

26. WHO WILL INVESTIGATE?	a. Installation Level Accident Investigation (IAI) Board Appointed ☐ Yes ☐ No	b. CAI Team Dispatched ☐ Yes ☐ No Team:

DA FORM 7305-R, APR 94

Figure A001-1. DA Form 7305-R, APR 94

WORKSHEET FOR
TELEPHONIC NOTIFICATION OF GROUND ACCIDENT
For use of this form, see AR 385-40: the proponent agency is OCSA

Immediately notify USASC telephonically of all Class A and B accidents IAW AR 385-40, chapter 3. Phone numbers are: Commercial (205) 255-2660/4273/3410 or DSN 558-2660/4273/3410

SHADED BLOCKS ARE FOR USASC USE ONLY

A. ASMIS CASE NUMBER

B. TIME & DATE OPERATIONS RECEIVED REPORT — a. Year | b. Month | c. Day | d. Time (local)

1. POINT OF CONTACT FOR ACCIDENT INFORMATION — a. Name: John Jones

b. Duty ☐ Commander ☐ Safety Officer ☐ Other *(Specify)*

c. Phone Number — DSN: Commercial: 913-555-3478

2. ACCIDENT CLASSIFICATION — ☒ A ☐ B

3. TIME & DATE OF ACCIDENT — a. Year: 98 | b. Month: May | c. Day: 10 | d. Time (local):

4. PERIOD OF DAY — ☐ Day ☐ Night

5. ON/OFF DUTY — ☐ On-Duty ☐ Off-Duty

6. TYPE OF EQUIPMENT/ MATERIAL INVOLVED

7. UNIT: 3-33 IN Bn, 21 AR Div

8. MACOM: IV Corps

9. NIGHT VISION DEVICE IN USE — ☐ Yes ☒ No

10. EXACT ACCIDENT LOCATION: Intersection of US Hwy 88 and US Hwy 11

11. ON-POST/OFF-POST? — ☐ On-post ☐ Off-post

12. MILITARY INSTALLATION NEAREST ACCIDENT SITE: Camp Sherman, MO

CHECK "YES" or "NO" FOR QUESTIONS 13 THROUGH 17	Yes	No
13. EXPLOSIVE/HAZARDOUS/SENSITIVE MATERIALS INVOLVED?		X
14. IF YES TO #13, ARE THEY SECURE?		X
15. ACCIDENT SITE SECURED IAW DA PAM385-40?		X
16. HAS ACCIDENT SITE BEEN DISTURBED?		X
17. IF YES TO #16, WERE PHOTOS, ETC. MADE BEFORE DISTURBING THE SCENE?		X

19. PERSONNEL INVOLVED — a. No. of Personnel by Rank/Category: 1 Officer ____ WO ____ Enlisted ____ Army Civilian ____ Non-Army Civilian; b. Total No. of Personnel: 1

18. WEATHER CONDITIONS: Raining

20. INJURIES (Enter # of each): 1 Fatalities ____ Non-Fatal Injuries

As soon as possible, the following Additional information is required On all injured personnel; name, Personnel classification, degree of Injury and SSAN.

21. ACCIDENT SYNOPSIS (What Happened)

William Smith crashed his automobile into a traffic light at the intersection of US Hwy 88 and US Hwy 11. Cause of accident has not been determined.

22. NEWS MEDIA AWARE OF ACCIDENT — ☐ Yes ☐ No

23. NEAREST AIRFIELD — a. Nearest that can handle C-12 (4 ,000 ft. min.); b. Nearest commercial airfield

24. WHO WILL INVESTIGATE? — a. Installation Level Accident Investigation (IAI) Board Appointed ☐ Yes ☐ No; b. CAI Team Dispatched ☐ Yes ☐ No Team:

DA FORM 7306-R, APR 94

Figure A001-2. DA Form 7306-R, APR 94

TITLE: ACKNOWLEDGE MESSAGE [AKNLDG]
REPORT NUMBER: A005 {USMTF # F541}

GENERAL INSTRUCTIONS: Use (1) to acknowledge receipt of message and indicate planned or accomplished action or (2) to validate intelligence collection requirements. Transmittal of the message implies understanding of the received message. Reference: FM 11-1.

LINE 1 – DATE AND TIME ______________________________ (DTG)

LINE 2 – UNIT ______________________________ (Unit Making Report)

LINE 3 – ACKNOWLEDGE ______________________________ (Report Type)

LINE 4 – ORIGINATOR ______________________________ (Originator of Report Acknowledged)

LINE 5 – SERIAL ______________________________ (Serial Number or DTG of Report Acknowledged)

LINE 6 – RESPONSE ______________________________ (Response to Report Acknowledged ACKNOWLEDGE, WILCO)

LINE 7 – VALIDATION STATUS ______________________________ (Response to Intelligence Collection Requirement: VALID, NONVALID, VALID WITH CHANGE)

LINE 8 – NARRATIVE ______________________________ (Free Text for Additional Information Required for Clarification of Report)

LINE 9 – AUTHENTICATION ______________________________ (Report Authentication)

TITLE: AIR DEFENSE COMMAND MESSAGE [AIRDEFCOM]
REPORT NUMBER: A010 {USMTF # E710}

GENERAL INSTRUCTIONS: Use (1) to direct weapon system engagement for defense or air support and (2) to provide receipt of or compliance with commands. Reference: FM 44-100.

LINE 1 – DATE AND TIME ______________________________ (DTG)

LINE 2 – UNIT ______________________________ (Unit Making Report)

LINE 3 – COMMAND ______________________________ (Command Action Directed)

LINE 4 – TRACK ______________________________ (Track Number)

LINE 5 – POSITION ______________________________ (UTM or Six-Digit Grid Coordinate With MGRS Grid Zone Designator)

LINE 6 – NO. TYPE ______________________________ (Number and Type of Aircraft)

LINE 7 – WEAPON ______________________________ (Weapon Type)

LINE 8 – NARRATIVE ______________________________ (Free Text for Additional Information Required for Clarification of Report)

LINE 9 – AUTHENTICATION ______________________________ (Report Authentication)

TITLE: AIR MISSION REQUEST STATUS/TASKING [REQSTATASK]
REPORT NUMBER: A015 {USMTF # A661}

GENERAL INSTRUCTIONS: Use (1) to inform the requesting component command whether a request has been approved or disapproved and (2) to inform the complying command that it will satisfy the request in the immediate air tasking, SAR, or airlift. Reference: FM 1-100. (See also Figure A015.)

LINE 1 – DATE AND TIME ______________________ (DTG)

LINE 2 – UNIT ______________________ (Unit Making Report)

LINE 3 – REQUEST ______________________ (Request Number)

LINE 4 – COMMAND ______________________ (Requesting Command)

LINE 5 – STATUS ______________________ (Status of Request: APPROVED or DISAPPROVED)

** If status is DISAPPROVED, skip to Line 10.

LINE 6 – TASKED ______________________ (Command Tasked to Comply With and Satisfy the Request)

LINE 7 – PRIORITY ______________________ (Revised Priority: 1, 1A through 1Z; 2, 2A through 2Z; 3, 3A Through 3Z; 4, 4A through 4Z)

LINE 8 – MISSION ______________________ (Type of Mission: CAS, EW, RECONNAISSANCE, and so on)

LINE 9 – START ______________________ (DTG Time That Aircraft Are to Arrive on Mission Location)

LINE 10 – NO. TYPE ______________________ (Number and Type Aircraft)

LINE 11 – NARRATIVE ______________________ (Free Text for Additional Information Required for Clarification of Report)

LINE 12 – AUTHENTICATION ______________________ (Report Authentication)

SAMPLE TACAIR Briefing Form (9 Line)

(Omit data not required. Do not transmit line number. Units of measure are standard unless otherwise specified. *Denotes minimum essential information in limited communications. Bold denotes readback items when requested.)

Terminal controller.________________, this is________________.
(Aircraft Call Sign) (Terminal Controller)

Line 1 - ***IP/BP**: ________________.

Line 2 - *Heading:________________ (Magnetic).
(IP/BP to Target)

Offset: ________________(Left/Right).

Line 3 - *Distance: ________________.
(IP-to-target in nautical miles/ BP-to-target in meters)

Line 4 - *Target Elevation: ________________(in feet MSL).

Line 5 - *Target Description: ________________.

Line 6 - ***Target Location**: ________________.
(Latitude/Longitude or Grid Coordinates or Offsets or Visual)

Line 7 - Type Mark: ________________Code:________________.
(Laser, WP, IR, Beacon) (Actual Code)

Line 8 - ***Location of Friendlies**: ________________.

Position Marked by: ________________.

Line 9 - Egress: ________________.

In the event of a beacon request, insert beacon bombing chart line numbers below.

________________.

Remarks (As Appropriate):________________.
(Threats, reactions, danger close, attack clearances, SEAD, abort codes, hazards)

NOTE: For AC-130 employment, lines 5, 6, and 8 are mandatory briefing items. Remarks should also include detailed threat description, marking method of friendly locations (including magnetic bearing and distance in meters from the friendly position to the target, if available). Identifiable ground features, danger close acceptance.

Time on Target (TOT):________________.

Or

Time to Target (TTT): Standby________________plus________________Hack.

Figure A015. Sample TACAIR Briefing Form (9 Line)

TITLE: AIR SUPPORT REQUEST [AIRSUPREQ]
REPORT NUMBER: A020 {USMTF # D670}

GENERAL INSTRUCTIONS: Use to request preplanned and immediate close-air support, interdiction, reconnaissance, surveillance, escort, helicopter airlift, and other aircraft missions. For expedient hard copy or voice request use DD Form 1972 (Revised) 15 NOV 1994 (see Figure A020) or Sample TACAIR Briefing Form (see Figure A015).

LINE 1 – DATE AND TIME ________________________ (DTG)

LINE 2 – UNIT ________________________ (Unit Making Request)

LINE 3 – EXER ________________________ (Exercise ID)

LINE 4 – OPER________________________ (Operation ID Data)

LINE 5 – MSGID ________________________ (AIRSUPREQ)

LINE 6 – REF ________________________ (Reference)

LINE 7 – AMPM________________________ (Amplification)

LINE 8 – NARRATIVE ________________________ (Free Text for Additional Information Required for Clarification of Report)

LINE 9 – CANX ________________________ (Message Cancellation With New Information Provided)

LINE 10 – PERID________________________ (Effective DTG of Requested Mission)

LINE 11 – REQUEST________________________ (Air Mission Request Information)

LINE 12 – MSNLOC ________________________ (Air Mission Location Information)

LINE 13 – CONTROL ________________________ (Air Mission Control Agency Information)

LINE 14 – GNDFRNF ________________________ (Ground Location Information of Friendly Troops)

LINE 15 – TGTDESC________________________ (Target Description Information)

**Continued on next page.

TITLE: AIR SUPPORT REQUEST [AIRSUPREQ] (CONTINUED)
REPORT NUMBER: A020 {USMTF # D670}

LINE 16 – GUIDED ______________________ (Laser Guided Weapons Information)

LINE 17 – BEACON______________________ (Beacon Description)

LINE 18 – SHIPTGT ______________________ (Ship Target Information)

LINE 19 – SHIPPOS______________________ (Ship Position and Movement Data)

LINE 20 – MOVAA ______________________ (Movement and Antiair Warfare Axis Information)

LINE 21 – FACSCD______________________ (Forward Air Controller Aircraft Schedule)

LINE 22 – EWDATA ______________________ (Electronic Warfare Information)

LINE 23 – CHAFE ______________________ (Chafe Mission Information)

LINE 24 – RECDATA ______________________ (Reconnaissance Mission Information)

LINE 25 – DELADDR ______________________ (Delivery Address)

LINE 26 – REPORT ______________________ (Report Requirements)

LINE 27 – AIRDROP ______________________ (Drop Zone Data)

LINE 28 – ROUTE ______________________ (Aircraft Route Data)

LINE 29 – REFUELD______________________ (Air-to-Air Refueling Data)

LINE 30 – REFUEL ______________________ (Air-to-Air Refueling)

LINE 31 – ESCDATA______________________ (Escort Mission Data)

LINE 32 – AUTHENTICATION ______________________ (Report Authentication)

JOINT TACTICAL AIR STRIKE REQUEST	See Joint Pub 3-09.3 for preparation instructions.

SECTION I - MISSION REQUEST

DATE

1. UNIT CALLED	THIS IS	REQUEST NUMBER	SENT
W7T35	V3B21		TIME / BY

2. PREPLANNED: [A] PRECEDENCE ______ [B] PRIORITY ______
IMMEDIATE: [C] PRIORITY 3

RECEIVED — TIME / BY

3. [A] PERS IN OPEN ______ [B] PERS DUG IN ______ [C] WPNS/MGT/RR/AT ______ [D] MORTARS, ARTY ______
[E] AAA ADA ______ [F] RKTS MISSILE ______ [G] ARMOR ______ [H] VEHICLES ______
[I] BLDGS ______ [J] BRIDGES 1 [K] PILLBOX,BUNKERS ______ [L] SUPPLIES, EQUIP ______
[M] CENTER (CP,COM) ______ [N] AREA ______ [O] ROUTE ______ [P] MOVING N E S W ______
[Q] REMARKS ______

4. TARGET LOCATION IS
[A] UW35847211 (COORDINATES) [B] ______ (COORDINATES) [C] ______ (COORDINATES) [D] ______ (COORDINATES)
[E] TGT ELEV ______ [F] SHEET NO. ______ [G] SERIES ______ [H] CHART NO. ______

5. TARGET TIME / DATE
[A] ASAP X [B] NLT ______ [C] AT ______ [D] TO ______

6. DESIRED ORD/ RESULTS [A] ORDNANCE ______
[B] DESTROY X [C] NEUTRALIZE ______ [D] HARASS/INTERDICT ______

7. FINAL CONTROL
[A] FAC/RABFAC FAC [B] CALL SIGN V3B21 [C] FREQ ______
[D] CONT PT 900 ft

8. REMARKS
1. IP ______
2. HDNG ______ MAG OFFSET: L/R
3. DISTANCE ______
4. TGT ELEVATION ______
5. TGT DESCRIPTION ______
6. TGT LOCATION ______
7. MARK TYPE ______ CODE ______
8. FRIENDLIES ______
9. EGRESS ______
10. BCN-TGT ______ MAG BCN GRID ______
11. BCN-TGT ______ METERS TGT GRID ______
12. BCN ELEVATION ______

SECTION II - COORDINATION

9. NGF	10. ARTY	11. AIO/G-2/G-3
12. REQUEST [ ] APPROVED [ ] DISAPPROVED	13. BY	14. REASON FOR DISAPPROVAL

15. AIRSPACE COORDINATION AREA [A] IS NOT IN EFFECT [B] NUMBER
16. IS IN EFFECT [A] (FROM TIME) ______ [B] (TO TIME) ______
17. LOCATION [A] ______ [B] ______
18. WIDTH (METERS)
19. ALTITUDE/VERTEX [A] ______ (MAXIMUM/VERTEX) [B] ______ MINIMUM

SECTION III - MISSION DATA

20. MISSION NUMBER	21. CALL SIGN	22. NO. AND TYPE AIRCRAFT	23. ORDNANCE
24. EST/ACT TAKEOFF	25. EST TOT	26. CONT PT(COORDS)	27. INITIAL CONTACT
28. FAC/FAC(A)/TAC(A)CALL SIGN FREQ	29. AIRSPACE COORDINATION AREA	30. TGT DESCRIPTION	*31. TGT COORD/ELEV

32. BATTLE DAMAGE ASSESSMENT (BDA) REPORT (USMTF INFLTREP)
LINE 1 / CALL SIGN ______ LINE 4 / LOCATION ______
LINE 2 / MSN NUMBER ______ LINE 5 / TOT ______
LINE 3 / REQ NUMBER ______ LINE 6 / RESULTS ______
REMARKS ______

*TRANSMIT AS APPROPRIATE.

DD FORM 1972 (REVISED) 15 NOV 1994. Supersedes DD Form 1972, Apr 1975.

Figure A020. DD Form 1972, Joint Tactical Air Strike Request

TITLE: AIRSPACE CONTROL MEANS REQUEST [ACMREQ]
REPORT NUMBER: A025 {USMTF # F658}

GENERAL INSTRUCTIONS: Use to request that a specific airspace control means be specified in a future airspace control order. Reference: FM 100-103.

LINE 1 – DATE AND TIME ______________________ (DTG)

LINE 2 – UNIT ______________________ (Unit Making Report)

LINE 3 – REQUESTING UNIT ______________________ (Name of Unit or Agency Requesting Airspace)

LINE 4 – TYPE ______________________ (Type of Airspace)

LINE 5 – AIRSPACE ______________________ (Name of Airspace)

LINE 6 – PURPOSE ______________________ (Purpose of Special Use Airspace)

LINE 7 – TRANSIT ______________________ (Transit Instructions)

LINE 8 – FROM ______________________ (DTG to be Established)

LINE 9 – UNTIL ______________________ (DTG to be Disestablished)

LINE 10 – AREA ______________________ (Description of the Area to be Defined (Boundary or Circle))

LINE 11 – COORDINATES ______________________ (UTM or Six-Digit Grid Coordinate With MGRS Grid Zone Designator)

LINE 12 – WIDTH/RADIUS ______________________ (Width or Radius)

LINE 13 – LOWER ______________________ (Lower Altitude of Designated Area to Nearest 100 Feet or Ground Level)

LINE 14 – UPPER ______________________ (Upper Altitude of Designated Area to Nearest 100 Feet)

LINE 15 – CONTACT ______________________ (Call Sign of Control Agency)

LINE 16 – PRIMARY ______________________ (Primary Frequency or Frequency Designator)

**Continued on next page.

TITLE: AIRSPACE CONTROL MEANS REQUEST [ACMREQ] (CONTINUED)
REPORT NUMBER: A025 {USMTF # F658}

LINE 17 – SECONDARY ______________________________ (Secondary Frequency or Frequency Designator)

LINE 18 – NARRATIVE ______________________________ (Free Text for Additional Information Required for Clarification of Report)

LINE 19 – AUTHENTICATION ______________________________ (Report Authentication)

TITLE: AIRSPACE CONTROL ORDER [ACO]
REPORT NUMBER: A030 {USMTF # F756}

GENERAL INSTRUCTIONS: Use to provide specific detailed orders for airspace management from a higher command to subordinate units. Reference: FM 100-103-2.

LINE 1 – DATE AND TIME ________	(DTG)
LINE 2 – UNIT ________	(Unit Making Report)
LINE 3 – AIRSPACE ________	(Type of Airspace)
LINE 4 – NUMBER ________	(Name or Serial Number of Area)
LINE 5 – FROM ________	(DTG Area to be Established)
LINE 6 – UNTIL ________	(DTG Area to be Disestablished)
LINE 7 – ACTION ________	(Type Action: ESTABLISH or CANCEL)
LINE 8 – SERIAL NUMBER ________	(ACO Serial Number)
LINE 9 – AREA ________	(Description of the Area to be Defined (Boundary or Circle))
LINE 10 – COORDINATES ________	(UTM or Six-Digit Grid Coordinate With MGRS Grid Zone Designator)
LINE 11 – WIDTH/RADIUS ________	(Width or Radius)
LINE 12 – LOWER ________	(Lower Altitude of Designated Area to Nearest 100 Feet or Ground Level)
LINE 13 – UPPER ________	(Upper Altitude of Designated Area to Nearest 100 Feet)
LINE 14 – CONTACT ________	(Call Sign of Control Agency)
LINE 15 – PRIMARY ________	(Primary Frequency or Frequency Designator)
LINE 16 – SECONDARY ________	(Secondary Frequency or Frequency Designator)

**Continued on next page.

TITLE: AIRSPACE CONTROL ORDER [ACO] (CONTINUED)
REPORT NUMBER: A030 {USMTF # F756}

LINE 17 – NARRATIVE	__________________________________	(Free Text for Additional Information Required for Clarification of Report)
LINE 18 – AUTHENTICATION	______________________________	(Report Authentication)

TITLE: AMMUNITION FIRE UNIT-AMMUNITION STATUS [AFU.AMS]
REPORT NUMBER: A035 {USMTF # B230}

GENERAL INSTRUCTIONS: Use to report ammunition status periodically or when requested.
Reference: FM 6-20-30 and FM 6-20-40.

LINE 1 – DATE AND TIME ______________________ (DTG)

LINE 2 – UNIT ______________________ (Unit Making Report)

LINE 3 – PLAN ______________________ (Fire Plan Designation)

LINE 4 – AMMO ______________________ (Projectile Type and Quantity)

LINE 5 – AMMO ______________________ (Projectile Type and Quantity)

LINE 6 – AMMO ______________________ (Projectile Type and Quantity)

LINE 7 – AMMO ______________________ (Projectile Type and Quantity)

LINE 8 – AMMO ______________________ (Projectile Type and Quantity)

LINE 9 – AMMO ______________________ (Projectile Type and Quantity)

LINE 10 – AMMO ______________________ (Propellant Type and Quantity)

LINE 11 – AMMO ______________________ (Propellant Type and Quantity)

LINE 12 – AMMO ______________________ (Propellant Type and Quantity)

LINE 13 – AMMO ______________________ (Propellant Type and Quantity)

LINE 14 – AMMO ______________________ (Propellant Type and Quantity)

LINE 15 – AMMO ______________________ (Propellant Type and Quantity)

LINE 16 – AMMO ______________________ (Propellant Type and Quantity)

LINE 17 – FUZE ______________________ (Fuze Type and Quantity)

LINE 18 – FUZE ______________________ (Fuze Type and Quantity)

LINE 19 – FUZE ______________________ (Fuze Type and Quantity)

LINE 20 – FUZE ______________________ (Fuze Type and Quantity)

LINE 21 – FUZE ______________________ (Fuze Type and Quantity)

**Continued on next page.

TITLE: AMMUNITION FIRE UNIT-AMMUNITION STATUS [AFU.AMS] (CONTINUED)
REPORT NUMBER: A035 {USMTF # B230}

LINE 22 – WARHEAD ______________________	(Warhead Mark Number, Nuclear Weapon Model Y Number)
LINE 23 – NUCLEAR WEAPON QTY ______________________	(Nuclear Weapons Quantity)
LINE 24 – NUCLEAR WEAPON QTY ______________________	(Nuclear Weapons Quantity)
LINE 25 – NARRATIVE ______________________	(Free Text for Additional Information Required for Clarification of Report)
LINE 26 – AUTHENTICATION ______________________	(Report Authentication)

TITLE: AMMUNITION FIRE UNIT-AMMUNITION SUPPLY RATE [AFU.ASR]
REPORT NUMBER: A040 {USMTF # F231}

GENERAL INSTRUCTIONS: Use (1) to report the required supply rate, (2) to establish the controlled supply rate, and (3) to update the amount of expended ammunition in relation to the controlled supply rate. Reference: FM 6-20-30 and FM 6-20-40.

LINE 1 – DATE AND TIME ______________________________ (DTG)

LINE 2 – UNIT ______________________________ (Unit Making Report)

LINE 3 – POI ______________________________ (Primary Option Indicator. Enter ADD, AMEND, or CANCEL)

LINE 4 – PLAN ______________________________ (Fire Plan Designation)

LINE 5 – WEAPON ______________________________ (Surface-to-Surface Weapon Type)

LINE 6 – REQUIRED______________________________ (Projectile Type and Required Supply Rate)

LINE 7 – REQUIRED______________________________ (Projectile Type and Required Supply Rate)

LINE 8 – REQUIRED______________________________ (Projectile Type and Required Supply Rate)

LINE 9 – REQUIRED______________________________ (Projectile Type and Required Supply Rate)

LINE 10 – REQUIRED______________________________ (Projectile Type and Required Supply Rate)

LINE 11 – REQUIRED______________________________ (Projectile Type and Required Supply Rate)

LINE 12 – CONTROLLED ______________________________ (Projectile Type and Controlled Supply Rate)

LINE 13 – CONTROLLED ______________________________ (Projectile Type and Controlled Supply Rate)

LINE 14 – CONTROLLED ______________________________ (Projectile Type and Controlled Supply Rate)

LINE 15 – CONTROLLED ______________________________ (Projectile Type and Controlled Supply Rate)

**Continued on next page.

TITLE: AMMUNITION FIRE UNIT-AMMUNITION SUPPLY RATE [AFU.ASR] (CONTINUED)
REPORT NUMBER: A040 {USMTF # F231}

LINE 16 – CONTROLLED ______________________________ (Projectile Type and Controlled Supply Rate)

LINE 17 – CONTROLLED ______________________________ (Projectile Type and Controlled Supply Rate)

LINE 18 – EXPENDED ______________________________ (Projectile Type and Ammunition Expended)

LINE 19 – EXPENDED ______________________________ (Projectile Type and Ammunition Expended)

LINE 20 – EXPENDED ______________________________ (Projectile Type and Ammunition Expended)

LINE 21 – EXPENDED ______________________________ (Projectile Type and Ammunition Expended)

LINE 22 – EXPENDED ______________________________ (Projectile Type and Ammunition Expended)

LINE 23 – EXPENDED ______________________________ (Projectile Type and Ammunition Expended)

LINE 24 – NARRATIVE ______________________________ (Free Text for Additional Information Required for Clarification of Report)

LINE 25 – AUTHENTICATION ______________________________ (Report Authentication)

TITLE: AMMUNITION FIRE UNIT-DEPLOYMENT COMMAND [AFU.DCMD]
REPORT NUMBER: A045 {USMTF # A261}

GENERAL INSTRUCTIONS: Use to exchange battalion/regiment missions and assignments and changes thereto, preplanned fire position areas, and movement orders for battalion/regiment or fire units. Reference: FM 6-20-30 and FM 6-20-40.

LINE 1 – DATE AND TIME ______________	(DTG)
LINE 2 – UNIT ______________	(Unit Making Report)
LINE 3 – POI ______________	(Primary Option Indicator. Enter ADD, AMEND, or CANCEL)
LINE 4 – PLAN ______________	(Fire Plan Designation)
LINE 5 – WEAPON ______________	(Surface-to-Surface Weapon Type)
LINE 6 – MISSION ______________	(Mission of the Fire Unit)
LINE 7 – ZOR______________	(Zone of Responsibility of the Fire Unit)
LINE 8 – TIME OF MISSION ______________	(DTG the Mission Assignment is Effective)
LINE 9 – TIME MISSION ENDS ______________	(DTG the Mission Assignment End)
LINE 10 – SUPPORT DESIGNATOR______________	(Use When the Mission Assigned in Line 6 is DS, GS, or GSR. Enter the Battery/Company Designator (if Appropriate) Followed by the Battalion Designator; Followed by the Brigade/Division Designator of the Supported Unit; Followed by the Designator of the Corps or Echelon Above Corps Being Supported)
LINE 11 – UTM GZ ______________	(Enter the UTM Grid Zone and 100-KM Square Identification)

**Continued on next page.

TITLE: AMMUNITION FIRE UNIT-DEPLOYMENT COMMAND [AFU.DCMD] (CONTINUED)
REPORT NUMBER: A045 {USMTF # A261}

LINES 12 – 21 UTM__________	(Number of the Point Followed by the UTM 1-Meter Easting and Northing of Each Point Used to Delineate the Battery Position Area)
LINE 13 – AZ OF FIRE __________	(For FA Units, Enter the Azimuth of Fire of the Unit in Line 2. Not Used for (NSF) Units)
LINE 14 – DTG MVT OF FIRE UNIT __________	(DTG and Time Zone Movement the Fire Unit in Line 2 Will Begin)
LINE 15 – DTG MVT COMPLETE__________	(DTG and Time Zone the Fire Unit in Line 2 is to Close (Complete Movement) Into the New Position Area)
LINE 16 – NARRATIVE __________	(Free Text for Additional Information Required for Clarification of Report)
LINE 17 – AUTHENTICATION __________	(Report Authentication)

TITLE: AMMUNITION FIRE UNIT-FIRE STATUS [AFU.FUS]
REPORT NUMBER: A050 {USMTF # B220}

GENERAL INSTRUCTIONS: Use to provide information on the friendly fire unit situation for making tactical fire direction decisions. Reference: FM 6-20-30 and FM 6-20-40.

LINE 1 – DATE AND TIME ____________________	(DTG)
LINE 2 – UNIT ____________________	(Unit Making Report)
LINE 3 – POI ____________________	(Primary Option Indicator. Enter ADD, AMEND, or CANCEL)
LINE 4 – PLAN ____________________	(Fire Plan Designation)
LINE 5 – STATUS ____________________	(Fire Unit Status)
LINE 6 – RETURN____________________	(Expected DTG Return to Operation)
LINE 7 – WEAPONS ____________________	(Number of Weapons, Surface-to-Surface Weapon Type)
LINE 8 – MODEL NO.____________________	(Artillery Weapon Model Number)
LINE 9 – RESPONSE____________________	(Unit Response DTG in Minutes)
LINE 10 – ZONE ____________________	(UTM Grid Zone and 100-KM Square)
LINE 11 – GRID____________________	(UTM or Six-Digit Grid Coordinate With MGRS Grid Zone Designator)
LINE 12 – AZIMUTH ____________________	(Azimuth of Fire in Mils)
LINE 13 – MIN ____________________	(Minimum Range in Meters)
LINE 14 – MAX ____________________	(Munitions Range Limitation Designator and Maximum Range in Meters)

**Continued on next page.

TITLE: AMMUNITION FIRE UNIT-FIRE STATUS [AFU.FUS] (CONTINUED)
REPORT NUMBER: A050 {USMTF # B220}

LINE 15 – MAX ____________________ (Munitions Range Limitation Designator and Maximum Range in Meters)

LINE 16 – MAX ____________________ (Munitions Range Limitation Designator and Maximum Range in Meters)

LINE 17 – MAX ____________________ (Munitions Range Limitation Designator and Maximum Range in Meters)

LINE 18 – MAX ____________________ (Munitions Range Limitation Designator and Maximum Range in Meters)

LINE 19 – MAX ____________________ (Munitions Range Limitation Designator and Maximum Range in Meters)

LINE 20 – MAX ____________________ (Munitions Range Limitation Designator and Maximum Range in Meters)

LINE 21 – MAX RATE ____________________ (Maximum Rate of Fire in Rounds per Weapon per Minute)

LINE 22 – SUSTAIN ____________________ (Sustained Rate of Fire in Rounds per Weapon per Minute)

LINE 23 – AUTHORIZED ____________________ (Fire Unit Authorized Ammunition Indicator)

LINE 24 – AUTHORIZED ____________________ (Fire Unit Authorized Ammunition Indicator)

**Continued on next page.

TITLE: AMMUNITION FIRE UNIT-FIRE STATUS [AFU.FUS] (CONTINUED)
REPORT NUMBER: A050 {USMTF # B220}

LINE 25 – LEFT LIMIT ______________________ (Azimuth of Left Traverse Limit in Mils)

LINE 26 – RIGHT LIMIT ______________________ (Azimuth of Right Traverse Limit in Mils)

LINE 27 – MISSION ______________________ (Fire Unit Mission)

LINE 28 – ZONE OF FIRE ______________________ (Zone of Fire (Responsibility))

LINE 29 – SUPPORTING ______________________ (Battery/Company, Battalion/ Regiment, and Regiment/ Brigade/Division Designators/Corps or Echelon Above Corps)

LINE 30 – REINFORCING ______________________ (Battalion/Regiment and Regiment/Brigade/Division Designators)

LINE 31 – EXPOSURE ______________________ (Nuclear Radiation Exposure Status)

LINE 32 – NVC ______________________ (Nuclear Vulnerability Category)

LINE 33 – NARRATIVE ______________________ (Free Text for Additional Information Required for Clarification of Report)

**Repeat lines 1 through 33 if required. Assign sequential line numbers to succeeding iterations; for example, first iteration 1 through 33; second iteration, 1a through 33a, and so on.

LINE 34 – AUTHENTICATION ______________________ (Report Authentication)

TITLE: AMMUNITION FIRE UNIT-FIRING SITE DATA [AFU.FSD]
REPORT NUMBER: A055 {USMTF # B280}

GENERAL INSTRUCTIONS: Use to establish and exchange dispersed firing site data for individual cannon, rocket, or missile weapons for the purpose of making tactical fire direction decisions. Reference: FM 6-20-30 and FM 6-20-40.

LINE 1 – DATE AND TIME ______________________	(DTG)
LINE 2 – UNIT ______________________	(Unit Making Report)
LINE 3 – POI ______________________	(Primary Option Indicator. Enter ADD, AMEND, or CANCEL)
LINE 4 – PLAN ______________________	(Fire Plan Designation)
LINE 5 – ZONE ______________________	(UTM Grid Zone and 100-KM Square)
LINE 6 – POSITION______________________	(UTM or Six-Digit Grid Coordinate With MGRS Grid Zone Designator)
LINE 7 – UTM ______________________	(UTM 1-Meter Easting, UTM 1-Meter Northing, and Altitude in Meters)
LINE 8 – RESPONSE______________________	(Firing Site Response Time)
LINE 9 – HE______________________	(Number of High-Explosive Warheads)
LINE 10 – ICM______________________	(Number of Dual Purpose ICM Warheads)
LINE 11 – MISSILE ASSEMBLIES ______________________	(Number of Main Missile Assemblies)
LINE 12 – POSTURE______________________	(Firing Site Response Posture)
LINE 13 – STATUS ______________________	(Fire Unit Status)
LINE 14 – RETURN TO OPERATION______________________	(Expected Time, Return to Operation)
LINE 15 – RADIATION STATUS ______________________	(Nuclear Radiation Exposure Status)

**Continued on next page.

TITLE: AMMUNITION FIRE UNIT-FIRING SITE DATA [AFU.FSD] (CONTINUED)
REPORT NUMBER: A055 {USMTF # B280}

LINE 16 – NVC ______________________________ (Nuclear Vulnerability Category)

LINE 17 – WARHEAD UPLOADED ______________________________ (Warhead Type on Launcher)

LINE 18 – NARRATIVE ______________________________ (Free Text for Additional Information Required for Clarification of Report)

**Repeat lines 5 through 18 up to four iterations if required. Assign sequential line numbers to succeeding iterations; for example, first iteration, 5 through 18; second iteration, 19 through 32; third iteration, 33 through 46; and so on.

LINE 19 – AUTHENTICATION ______________________________ (Report Authentication)

TITLE: AMMUNITION FIRE UNIT-MISSION FIRED REPORT [AFU.MFR]
REPORT NUMBER: A060 {USMTF # C241}

GENERAL INSTRUCTIONS: Use to provide target information, ammunition expenditure, and target disposition following engagement of a target. Reference: FM 6-20-30, FM 6-20-40, and FM 6-71.

LINE 1 – DATE AND TIME ________________________ (DTG)

LINE 2 – UNIT ________________________ (Unit Making Report)

LINE 3 – TARGET NO. ________________________ (Fire Support Target Number)

LINE 4 – ZONE ________________________ (UTM Grid Zone and 100-KM Square)

LINE 5 – GRID ________________________ (UTM or Six-Digit Grid Coordinate With MGRS Grid Zone Designator)

LINE 6 – TARGET ELEMENTS ________________________ (Number of Target Elements)

LINE 7 – TARGET TYPE ________________________ (Target Type, Subtype, and Degree of Personnel Protection)

LINE 8 – RADIUS ________________________ (Target Radius in Meters)

LINE 9 – LENGTH ________________________ (Target Length in Meters)

LINE 10 – WIDTH ________________________ (Target Width in Meters)

LINE 11 – ATTITUDE ________________________ (Attitude in Mils)

LINE 12 – CREDIBILITY ________________________ (Target Acquisition Agency)

LINE 13 – TRA ________________________ (Target Report Accuracy)

LINE 14 – RELIABILITY ________________________ (Evaluation of Reliability)

LINE 15 – NO. TEC ________________________ (Number of Target Element Casualties)

LINE 16 – TARGET DISPOSITION ________________________ (Fire Support Target Disposition)

LINE 17 – SHIP CALL SIGN ________________________ (Ship Call Sign(s))

**Continued on next page.

TITLE: AMMUNITION FIRE UNIT-MISSION FIRED REPORT [AFU.MFR] (CONTINUED)
REPORT NUMBER: A060 {USMTF # C241}

LINE 18 – NO. RDS ______	(Number of Rounds Expended, Followed by Projectile Type, Followed by Fuze Type)
LINE 19 – UNIT ______	(Section, Platoon, and Battery/ Company, Battalion/Regiment, and Regiment/Brigade/Division Designators)
LINE 20 – NO. RDS ______	(Number of Rounds Expended, Followed by Projectile Type, Followed by Fuze Type)
LINE 21 – UNIT ______	(Section, Platoon, and Battery/ Company, Battalion/Regiment, and Regiment/Brigade/Division Designators)
LINE 22 – NO. RDS ______	(Number of Rounds Expended, Followed by Projectile Type, Followed by Fuze Type)
LINE 23 – NARRATIVE ______	(Free Text for Additional Information Required for Clarification of Report)
LINE 24 – AUTHENTICATION ______	(Report Authentication)

TITLE: ARTILLERY TARGET INTELLIGENCE-ARTILLERY TARGET REPORT [ATI.ATRI]
REPORT NUMBER: A070 {USMTF # C281}

GENERAL INSTRUCTIONS: Use to exchange artillery target information among fire support agencies. Reference: FM 6-20-30 and FM 6-20-40.

LINE 1 – DATE AND TIME ______________________ (DTG)

LINE 2 – UNIT ______________________ (Unit Making Report)

LINE 3 – POI ______________________ (Primary Option Indicator. Enter ADD, AMEND, or CANCEL)

LINE 4 – TARGET NO. ______________________ (Fire Support Target Number)

LINE 5 – EFFECTS ______________________ (Effects (Percent Damage) Required)

LINE 6 – ZONE ______________________ (UTM Grid Zone and 100-KM Square)

LINE 7 – GRID ______________________ (UTM or Six-Digit Grid Coordinate With MGRS Grid Zone Designator)

LINE 8 – TARGET ELEMENTS ______________________ (Number of Target Elements)

LINE 9 – TARGET ______________________ (Target Type and Subtype, and Degree of Personnel Protection)

LINE 10 – RADIUS ______________________ (Target Radius in Meters)

LINE 11 – LENGTH ______________________ (Target Length in Meters)

LINE 12 – WIDTH ______________________ (Target Width in Meters)

LINE 13 – ATTITUDE ______________________ (Attitude in Mils)

LINE 14 – CREDIBILITY ______________________ (Target Acquisition Agency)

LINE 15 – TARGET ACCURACY ______________________ (Target Report Accuracy in Meters)

LINE 16 – RELIABILITY ______________________ (Evaluation of Reliability)

LINE 17 – PERMANENCE ______________________ (Target Permanence)

**Continued on next page.

TITLE: ARTILLERY TARGET INTELLIGENCE-ARTILLERY TARGET REPORT [ATI.ATRI] (CONTINUED)
REPORT NUMBER: A070 {USMTF # C281}

LINE 18 – TARGET STATUS ______________________ (Mission Fired Indicator or Confirmed Target Indicator)

LINE 19 – NUCLEAR ______________________ (Target of Nuclear Interest Indicator)

LINE 20 – MASK ______________________ (Personnel Clothing and Mask Indicator)

LINE 21 – PROFICIENCY ______________________ (Enemy NBC Proficiency Level)

LINE 22 – VEGETATION ______________________ (Target Vegetation)

LINE 23 – CASUALTIES ______________________ (Number of Target Element Casualties)

LINE 24 – TGT DISPOSITION ______________________ (Fire Support Target Disposition)

LINE 25 – CATEGORY ______________________ (Information Request Category)

LINE 26 – SRI ______________________ (Standing Request for Artillery Target Information Number)

LINE 27 – NARRATIVE ______________________ (Free Text for Additional Information Required for Clarification of Report)

LINE 28 – AUTHENTICATION ______________________ (Report Authentication)

TITLE: ARTILLERY TARGET INTELLIGENCE-TARGET CRITERIA [ATI.TCRIT]
REPORT NUMBER: A075 {USMTF # D281}

GENERAL INSTRUCTIONS: Use to disseminate the commander's established targeting criteria. Reference: FM 6-20-30 and FM 6-20-40.

LINE 1 – DATE AND TIME ______________________ (DTG)

LINE 2 – UNIT ______________________ (Unit Making Report)

LINE 3 – POI ______________________ (Primary Option Indicator. Enter ADD, AMEND, or CANCEL)

LINE 4 – REQUEST ______________________ (Request Number, Target Criteria)

LINE 5 – ZONE ONE ______________________ (Zone of Fire (Responsibility))

LINE 6 – ZONE TWO ______________________ (Zone of Fire (Responsibility))
(OR)
LINE 7 – GRID ZONE ______________________ (UTM Grid Zone and 100-KM Square)

LINE 8 – QUADRANGLE ______________________ (UTM 1-Meter Easting and UTM 1-Meter Northing)

LINE 9 – UTM ______________________ (UTM 1-Meter Easting and UTM 1-Meter Northing)

LINE 10 – WIDTH ______________________ (Search Zone Width in Meters)
(OR)
LINE 11 – CIRCLE ______________________ (UTM 1-Meter Easting and UTM 1-Meter Northing)

LINE 12 – RADIUS ______________________ (Circular Radius in Meters)
(OR)
LINE 13 – FLOT DISTANCE ______________________ (Minimum Distance)

LINE 14 – MAX ______________________ (Maximum Distance)

LINE 15 – FORMAT ______________________ (Output Report Category)

LINE 16 – TARGET TYPE ______________________ (Target Type and Subtype)

**Continued on next page.

TITLE: ARTILLERY TARGET INTELLIGENCE-TARGET CRITERIA [ATI.TCRIT] (CONTINUED)
REPORT NUMBER: A075 {USMTF # D281}

LINE 17 – PROTECTION ______________________ (Degree of Personnel Protection)

LINE 18 – PERMANENCE ______________________ (Target Permanence)

LINE 19 – LOWER STRENGTH______________________ (Lower Strength Limit)

LINE 20 – UPPER STRENGTH ______________________ (Upper Strength Limit)

LINE 21 – LOWER SIZE______________________ (Lower Length Limit in Meters)

LINE 22 – LOWER WIDTH ______________________ (Lower Width Limit in Meters)

LINE 23 – UPPER SIZE______________________ (Upper Length Limit in Meters)

LINE 24 – UPPER WIDTH ______________________ (Upper Width Limit in Meters)

LINE 25 – ACCURACY______________________ (Target Report Accuracy in Meters)

LINE 26 – RELIABILITY ______________________ (Evaluation of Reliability)

LINE 27 – NARRATIVE ______________________ (Free Text for Additional Information Required for Clarification of Report)

LINE 28 – AUTHENTICATION ______________________ (Report Authentication)

TITLE: ASSET/MULTIPLE ASSET STATUS REPORT [ASTSTATREP]
REPORT NUMBER: A080

GENERAL INSTRUCTIONS: Use to convey the status of the unit's intelligence collectors. Send every four hours or upon loss to enemy or loss of contact with an asset. Send over the O&I net with PRIORITY precedence. Reference: FM 34-3.

LINE 1 – DATE AND TIME ________________________________ (DTG)

LINE 2 – UNIT ____ (Unit Making Report)

LINE 3 – T/O ____ (Task Organization of Intelligence Collectors)

LINE 4 – COLLECTOR __________________________________ (Collector by Team/System; Collectors by Type and Team/System)

LINE 5 – GRID ____ (UTM or Six-Digit Grid Coordinate With MGRS Grid Zone Designator or NO CHANGE)

LINE 6 – ACTIVITY ____________________________________ (Brief Summary of Activity)

LINE 7 – NARRATIVE __________________________________ (Free Text for Additional Information Required for Clarification of Report)

LINE 8 – AUTHENTICATION _____________________________ (Report Authentication)

TITLE: AVIATION (ARMY ROTARY WING) MISSION/SUPPORT REQUEST [AVIAREQ]
REPORT NUMBER: A085

GENERAL INSTRUCTIONS: Use to request Army aviation units for support. Send from subordinate units to division/corps aviation cell at the main command post. Reference: FM 1-100 and FM 1-114.

LINE 1 – DATE AND TIME ______________________ (DTG)

LINE 2 – UNIT ______________________ (Unit Making Report)

LINE 3 – ASSETS ______________________ (Assets Required)

EXAMPLE: Three (3) CH-47s

LINE 4 – PURPOSE ______________________ (Purpose)

EXAMPLE: Emergency Logistical Resupply of Class IV at BP 53

LINE 5 – PRIORITY ______________________ (Priority)

LINE 6 – DTG and GRID of PZ ______________________ (DTG and Location of Pickup Zone, include marking system, frequency and call sign of unit at the PZ)

LINE 7 – DTG and GRID of LZ ______________________ (DTG and Location of Loading Zone, include marking system, frequency and call sign of unit at the LZ)

LINE 8 – A2C2 ______________________ (A2C2 Route and Information (Ingress/Egress Route))

LINE 9 – ENEMY ADA ______________________ (Probability of Enemy ADA)

LINE 10 – TARGET ______________________ (Target Location and Information for Planning)

LINE 11 – LOAD ______________________ (Estimated Load Requirements (Slings/Type of Ammunition))

LINE 12 – POC ______________________ (POC of Requesting Unit for Coordination)

LINE 13 – POC AT PZ/LD ______________________ (POC at Pickup Zone or LD)

LINE 14 – POC AT LZ ______________________ (Unit Requesting POC at LZ)

**Continued on next page.

TITLE: AVIATION (ARMY ROTARY WING) MISSION/SUPPORT REQUEST [AVIAREQ] (CONTINUED)
REPORT NUMBER: A085

LINE 15 – FARP ______________________	(Suggested FARP Locations Within Supported Area)
LINE 16 – REMARKS ______________________	(Remarks with Safety Risk Assessment)
LINE 17 – WEATHER ______________________	(Aviation Weather Forecast Within Supported Areas)
LINE 18 – NARRATIVE ______________________	(Free Text for Additional Information Required for Clarification of Report)
LINE 19 – AUTHENTICATION ______________________	(Report Authentication)

TITLE: BASIC WIND DATA MESSAGE [BWD]
REPORT NUMBER: B001 {USMTF # C508}

GENERAL INSTRUCTIONS: Use to report wind direction and speed in 2,000 meter increments from the surface of the earth to an altitude of 30,000 meters. Reference: FM 34-81.

LINE 1 – DATE AND TIME ______________________________ (DTG)

LINE 2 – UNIT ______________________________ (Unit Making Report)

LINE 3 – TIME OF MEASUREMENT ______________________________ (DTG of Measurement)

LINE 4 – AREA ______________________________ (Area of Validity)

LINE 5 – LAYER MIKE ______________________________ (Layer Indicator, Downwind Direction in Degrees True, and Wind Speed in Kilometers per Hour or Knots)

LINE 6 – NARRATIVE ______________________________ (Free Text for Additional Information Required for Clarification of Report)

LINE 7 – AUTHENTICATION ______________________________ (Report Authentication)

TITLE: BATTLE DAMAGE ASSESSMENT REPORT [BDAREP]
REPORT NUMBER: B005 {USMTF # C104}

GENERAL INSTRUCTIONS: Use to provide a timely and accurate estimate of damage resulting from the application of military force, either lethal or non-lethal, against a predetermined objective. Reference: FM 9-43-1.

LINE 1 – DATE AND TIME ______________________	(DTG)
LINE 2 – UNIT ______________________	(Unit Making Report)
LINE 3 – NAME ______________________	(Target or Installation Name)
LINE 4 – COUNTRY CODE ______________________	(Two-Letter Country Code)
LINE 5 – BEN ______________________	(Basic Encyclopedia Number, NA or UNKNOWN)
LINE 6 – LOCATION ______________________	(UTM or Six-Digit Grid Coordinate With MGRS Grid Zone Designator)
LINE 7 – TM REF ______________________	(Target Material Reference)
LINE 8 – PAGE NO. ______________________	(Page or Sheet Number)
LINE 9 – COLLECTION DATE/TIME ______________________	(IMINT Collection DTG)
LINE 10 – TYPE ______________________	(Type of IMINT)
LINE 11 – QUALITY ______________________	(Image Quality – GOOD/FAIR/POOR)
LINE 12 – ANGLE ______________________	(Viewing Angle – VERTICAL or OBLIQUE)
LINE 13 – TOT ______________________	(Time Over Target for Attack DTG)
LINE 14 – DELIVERY SYSTEM ______________________	(Number and Type of Delivery System)
LINE 15 – WEAPONS ______________________	(Number and Type of Weapons and Fusing)
LINE 16 – AIM POINT NAME ______________________	(Name or Description of Target Element)

**Continued on next page.

TITLE: BATTLE DAMAGE ASSESSMENT REPORT [BDAREP] (CONTINUED)
REPORT NUMBER: B005 {USMTF # C104}

LINE 17 – LOCATION ______________________ (UTM or Six-Digit Grid Coordinate With MGRS Grid Zone Designator)

**Repeat lines 13 through 17 for multiple missions or sets of mission data. Assign sequential lines to succeeding iterations; for example, first iteration 13 through 17; second iteration 13a through 17a; third iteration 13b through 17b; and so on.

LINE 18 – BDA ANALYSIS ______________________ (Narrative of Target Damage and Munitions Effects to Include: Element Name, Grid Reference, Physical Damage, Confidence Level, and Whether Reattack is Required)

LINE 19 – NARRATIVE ______________________ (Free Text for Additional Information Required for Clarification of Report)

LINE 20 – AUTHENTICATION ______________________ (Report Authentication)

TITLE: BED AVAILABILITY AND ELEMENT STATUS [BEDAVAIL]
REPORT NUMBER: B010 {USMTF # B913}

GENERAL INSTRUCTIONS: Use to report bed availability and medical treatment element (MTE) status when MEDSITREP record message is not or can not be used. Reference: FM 8-10-5.

LINE 1 – DATE AND TIME ________________________________ (DTG)

LINE 2 – UNIT ____ (Unit Making Report)

LINE 3 – ELEMENT _______________________________________ (Name or Designator of the Reporting Element)

LINE 4 – TOTAL OPERATIONAL____________________________ (Total Number of Beds That Are Operational)

LINE 5 – MEDICAL OPERATIONAL __________________________ (Number of Medical (MM) Beds Operational)

LINE 6 – PSYCHIATRIC OPERATIONAL _______________________ (Number of Psychiatric (MP) Beds Operational)

LINE 7 – GENERAL OPERATIONAL __________________________ (Number of General Surgeries (SS) Beds Operational)

LINE 8 – ORTHOPEDIC OPERATIONAL_______________________ (Number of Orthopedic (SO) Beds Operational)

LINE 9 – BURN OPERATIONAL______________________________ (Number of Burn (SB) Beds Operational)

LINE 10 – SPINAL CORD OPERATIONAL ____________________ (Number of Spinal Cord (SC) Injury Beds Operational)

LINE 11 – OB/GYN OPERATIONAL___________________________ (Number of OB/GYN (SG) Beds Operational)

LINE 12 – PEDIATRIC OPERATIONAL_________________________ (Number of Pediatric (MC) Beds Operational)

LINE 13 – NEURO OPERATIONAL ____________________________ (Number of Neurosurgery (SSN) Beds Operational)

LINE 14 – ORAL OPERATIONAL______________________________ (Number of Oral/Maxillofacial (SSM) Operational)

**Continued on next page.

TITLE: BED AVAILABILITY AND ELEMENT STATUS [BEDAVAIL] (CONTINUED)
REPORT NUMBER: B010 {USMTF # B913}

LINE 15 – OPHTHALMOLOGY OPERATIONAL _______________ (Number of Ophthalmology (SSO) Beds Operational)

LINE 16 – THORACIC OPERATIONAL _______________ (Number of Thoracic Surgery (SSC) Beds Operational)

**Use lines 13 through 16 to report the number of operational beds in selected surgical sub-specialties. Include these numbers in line 6.

LINE 17 – AVAILABLE TOTAL _______________ (Total Number of Available Beds)

LINE 18 – AVAILABLE MEDICAL _______________ (Number of Medical (MM) Beds Available)

LINE 19 – AVAILABLE PSYCHIATRIC _______________ (Number of Psychiatric (MP) Beds Available)

LINE 20 – AVAILABLE GENERAL _______________ (Number of General Surgery (SS) Beds Available)

LINE 21 – AVAILABLE ORTHOPEDIC _______________ (Number of Orthopedic (SO) Beds Available)

LINE 22 – AVAILABLE BURN _______________ (Number of Burn (SB) Beds Available)

LINE 23 – AVAILABLE SPINAL CORD _______________ (Number of Spinal Cord (SC) Injury Beds Available)

LINE 24 – AVAILABLE OB/GYN _______________ (Number of OB/GYN (SG) Beds Available)

LINE 25 – AVAILABLE PEDIATRIC _______________ (Number of Pediatric (MC) Beds Available)

**Use lines 17 through 25 to report the number of available beds for patients at the reporting medical element.

LINE 26 – AVAILABLE NEURO _______________ (Number of Neurosurgery (SSN) Available)

LINE 27 – AVAILABLE ORAL _______________ (Number of Oral/Maxillofacial (SSM) Beds Available)

**Continued on next page.

TITLE: BED AVAILABILITY AND ELEMENT STATUS [BEDAVAIL] (CONTINUED)
REPORT NUMBER: B010 {USMTF # B913}

LINE 28 – AVAILABLE OPHTHALMOLOGY__________________ (Number of Ophthalmology (SSO) Beds Available)

LINE 29 – AVAILABLE THORACIC __________________________ (Number of Thoracic Surgery (SSC) Beds Available)

**Use lines 26 through 29 to report the number of available beds in selected surgical sub-specialties. Include these numbers in line 19.

LINE 30 – SUITES____ (Number of Operating Suites That Are Operational)

LINE 31 – BACKLOG _______________________________________ (Number of Hours Surgical Backlog)

LINE 32 – OVERFLOW _____________________________________ (Number of Overflow/Holding Beds Available)

LINE 33 – NARRATIVE _____________________________________ (Free Text for Additional Information Required for Clarification of Report)

**Use lines 30 through 33 to provide additional information concerning medical element status, workload, and other amplifying data.

**Repeat lines 3 through 33 for multiple missions or sets of mission data. Assign sequential line numbers to succeeding iterations; for example, first iteration 11 through 15; second iteration 11a through 15a; third iteration 11b through 15b; and so on.

LINE 34 – AUTHENTICATION ______________________________ (Report Authentication)

TITLE: BED DESIGNATIONS [BEDDESIG]
REPORT NUMBER: B015 {USMTF # A913}

GENERAL INSTRUCTIONS: Use to report bed availability and medical treatment element (MTE) status when the MEDREGREP record message can not be used. Reference: FM 8-10-5.

LINE 1 – DATE AND TIME ______________________	(DTG)
LINE 2 – UNIT ______________________	(Unit Making Report)
LINE 3 – EVAC FROM ______________________	(Name or Designator of MTE From Which to Move/Evacuate Patients)
LINE 4 – PICKUP ______________________	(UTM or Six-Digit Grid Coordinate With MGRS Grid Zone Designator of PICKUP or Location of MTE From Which Evacuated. Report Only if Other Than Fixed-Wing Aeromedical Evacuation (AE) Will be Used and Transporting Agency May Not Know the PICKUP Location)
LINE 5 – EVAC TO ______________________	(Name or Designator of Destination MTE)
LINE 6 – DELIVERY ______________________	(Delivery Location or Location of Destination MTE, UTM or Six-Digit Grid Coordinate With MGRS Grid Zone Designator)
LINE 7 – NO. MEDICAL (MM) ______________________	(Number of MM Beds Assigned to Patients at Destination MTE)
LINE 8 – NO. PSYCHIATRIC (MP) ______________________	(Number of MP Beds Assigned to Patients at Destination MTE)
LINE 9 – NO. SURGERY (SS) ______________________	(Number of SS Beds Assigned to Patients at Destination MTE)
LINE 10 – NO. ORTHOPEDIC (SO) ______________________	(Number of SO Beds Assigned to Patients at Destination MTE)
LINE 11 – NO. BURN (SB) ______________________	(Number of SB Beds Assigned to Patients at Destination MTE)

**Continued on next page.

TITLE: BED DESIGNATIONS [BEDDESIG] (CONTINUED)
REPORT NUMBER: B015 {USMTF # A913}

LINE 12 – NO. SPINAL CORD (SC) ____________	(Number of SC Beds Assigned to Patients at Destination MTE)
LINE 13 – NO. OB/GYN (SG) ____________	(Number of SG Beds Assigned to Patients at Destination MTE)
LINE 14 – NO. PEDIATRIC (MC) ____________	(Number of MC Beds Assigned to Patients at Destination MTE)
LINE 15 – TOTAL ____________	(Total Number of Beds Assigned to Patients at Destination MTE)
LINE 16 – NO. NEURO (SSN) ____________	(Number of SSN Beds Assigned to Patients at Destination MTE)
LINE 17 – NO. MAXILLO (SSM) ____________	(Number of SSM Beds Assigned to Patients at Destination MTE)
LINE 18 – NO. OPHTHALMOLOGY (SSO) ____________	(Number of SSO Beds Assigned to Patients at Destination MTE)
LINE 19 – NO. THORACIC (SSC) ____________	(Number of SSC Beds Assigned to Patients at Destination MTE)

**Use lines 16 through 19 to report the number of beds assigned to patients in selected surgical subspecialties. Include these numbers in line 8.

LINE 20 – CATEGORY ____________	(Number of Special Category Patients (Other Than Active Duty US Armed Forces) Assigned Beds at Destination MTE by Medical Specialty Category (EXAMPLE: 2 French Nationals in Category SO. The Number of Patients Reported is Also Included in Lines 6 to 14))
LINE 21 – NARRATIVE ____________	(Free Text for Additional Information Required for Clarification of Report)

**Continued on next page.

TITLE: BED DESIGNATIONS [BEDDESIG] (CONTINUED)
REPORT NUMBER: B015 {USMTF # A913}

**Repeat lines 3 through 21 for multiple missions or sets of mission data. Assign sequential line numbers to succeeding iterations; for example, first iteration 3 through 21; second iteration 3a through 21a; third iteration 3b through 21b; and so on.

LINE 22 – AUTHENTICATION ______________________________ (Report Authentication)

**Use lines 20 through 22 to provide additional medical information and other amplifying data.

TITLE: BED REQUEST [BEDREQ]
REPORT NUMBER: B020 {USMTF # C913}

GENERAL INSTRUCTIONS: Use to request beds for patients when the medical group record can not be used. Reference: FM 8-10-5.

LINE 1 – DATE AND TIME ______________________________	(DTG)
LINE 2 – UNIT ______________________________	(Unit Making Report)
LINE 3 – REQUESTOR ______________________________	(Name or Designator of Requesting Element With Patients Requiring Beds)
LINE 4 – LOCATION ______________________________	(UTM or Six-Digit Grid Coordinate With MGRS Grid Zone Designator. Report Only on First Report or Upon Relocation)
LINE 5 – PATIENTS ___________HYPHEN _____ TOTAL ______	(Number of Litter (After PATIENTS), Ambulatory (After HYPHEN), and All (After TOTAL) Patients Requiring Beds)
LINE 6 – MEDICAL (MM) _______HYPHEN _____ TOTAL ______	(Number of Litter (After MM), Ambulatory (After HYPHEN), and All (After TOTAL) MM Requiring Beds)
LINE 7 – PSYCHIATRIC (MP)____HYPHEN _____ TOTAL ______	(Number of Litter (After MP), Ambulatory (After HYPHEN), and All (After TOTAL) MP Requiring Beds)
LINE 8 – SURGERY (SS) ________HYPHEN _____ TOTAL ______	(Number of Litter (After SS), Ambulatory (After HYPHEN), and All (After TOTAL) SS Requiring Beds)
LINE 9 – ORTHOPEDIC (SO) ____HYPHEN _____ TOTAL ______	(Number of Litter (After SO), Ambulatory (After HYPHEN), and All (After TOTAL) SO Requiring Beds)

**Continued on next page.

TITLE: BED REQUEST [BEDREQ] (CONTINUED)
REPORT NUMBER: B020 {USMTF # C913}

LINE 10 – BURN (SB) _________	HYPHEN _____	TOTAL ______	(Number of Litter (After SB), Ambulatory (After HYPHEN), and All (After TOTAL) SB Requiring Beds)
LINE 11 – SPINAL CORD (SC) ___	HYPHEN _____	TOTAL ______	(Number of Litter (After SC), Ambulatory (After HYPHEN), and All (After TOTAL) SC Requiring Beds)
LINE 12 – OB/GYN (SG) ________	HYPHEN _____	TOTAL ______	(Number of Litter (After SG), Ambulatory (After HYPHEN), and All (After TOTAL) SG Requiring Beds)
LINE 13 – PEDIATRIC (MC) _____	HYPHEN _____	TOTAL ______	(Number of Litter (After MC), Ambulatory (After HYPHEN), and All (After TOTAL) MC Requiring Beds)
LINE 14 – NEURO (SSN) ________	HYPHEN _____	TOTAL ______	(Number of Litter (After SSN), Ambulatory (After HYPHEN), and All (After TOTAL) SSN Requiring Beds)
LINE 15 – MAXILLO (SSM) _____	HYPHEN _____	TOTAL ______	(Number of Litter (After SSM), Ambulatory (After HYPHEN), and All (After TOTAL) SSM Requiring Beds)
LINE 16 – OPHTHAL (SSO) _____	HYPHEN _____	TOTAL ______	(Number of Litter (After SSO), Ambulatory (After HYPHEN), and All (After TOTAL) SSO Requiring Beds)
LINE 17 – THORACIC (SSC) _____	HYPHEN _____	TOTAL ______	(Number of Litter (After SSC), Ambulatory (After HYPHEN), and All (After TOTAL) SSC Requiring Beds)

**Use lines 14 through 17 to report the number of patients requiring beds in selected surgical sub-specialties. Include these numbers in line 8. Assign sequential line numbers to succeeding iterations; for example, first iteration 14 through 16; second iteration 14a through 17a; third iteration 14b through 17b; and so on.

**Continued on next page.

TITLE: BED REQUEST |BEDREQ| (CONTINUED)
REPORT NUMBER: B020 {USMTF # C913}

LINE 18 – CATEGORY ______________________________	(Number of Special Category Patients (Other Than Active Duty US Armed Forces) Assigned Beds at Destination MTE by Medical Specialty Category (EXAMPLE: 2 French Nationals))
LINE 19 – PICKUP ______________________________	(Pickup Location in UTM or Six-Digit Grid Coordinate With MGRS Grid Zone Designator)
LINE 20 – ONLOAD __________ HYPHEN _____ TOTAL _____	(Number of Litter (After ONLOAD), Ambulatory (After HYPHEN), and All (After TOTAL) Patients to be Picked Up)
LINE 21 – EQUIPMENT ______________________________	(Special Medical Equipment Required)
LINE 22 – NARRATIVE ______________________________	(Free Text for Additional Information Required for Clarification of Report)

**Repeat lines 19-22 as a group to identify patient pickup location and time available for movement when involving more than one location and time. Assign sequential line numbers to succeeding iterations; for example, first iteration 19 through 21; second iteration 19a through 21a; third iteration 19b through 21b; and so on.

**Repeat lines 3 through 22 as a group for each MTE requesting beds when reporting for more than one MTE. Assign sequential line numbers to succeeding iterations; for example, first iteration 2 through 22; second iteration 2a through 22a; third iteration 2b through 22b; and so on.

LINE 23 – AUTHENTICATION ______________________________	(Report Authentication)

TITLE: BLOOD SHIPMENT REPORT [BLDSHIPREP]
REPORT NUMBER: B025 {USMTF # B911}

GENERAL INSTRUCTIONS: Use to report blood shipments. Reference: FM 8-10-5.

LINE 1 – DATE AND TIME ______________________ (DTG)

LINE 2 – UNIT ______________________ (Unit Making Report)

LINE 3 – ACTIVITY ______________________ (Reporting Unit's Activity Code Letter:
A = Joint Blood Program Office
B= Area Joint Blood Program Office
C = Armed Services Whole Blood Processing Laboratory
D = Blood Donor Center
E = Blood Products Depot
F = Blood Transshipment Center
G = Blood Supply Unit
H = Medical Treatment Element
I = Naval Vessel)

LINE 4 – LOCATION ______________________ (UTM or Six-Digit Grid Coordinate With MGRS Grid Zone Designator)

LINE 5 – RENDEZVOUS ______________________ (NAVAL VESSELS ONLY: UTM or Six-Digit Grid Coordinate With MGRS Grid Zone Designator for Delivery of Blood Products)

LINE 6 – ARRIVAL ______________________ (NAVAL VESSELS ONLY (Hospital Ship): Estimated DTG at the Projected Location)

LINE 7 – PRODUCT ______________________ (Brevity Code Letter of Blood Product Being Shipped:
J =Red Blood Cells
K =Whole Blood
L =Frozen Red Blood Cells
M =Fresh Frozen Plasma
N =Frozen Platelets)

LINE 8 – O POSITIVE ______________________ (Number of Units)

**Continued on next page.

TITLE: BLOOD SHIPMENT REPORT [BLDSHIPREP] (CONTINUED)
REPORT NUMBER: B025 {USMTF # B911}

LINE 9 – O NEGATIVE____________________ (Number of Units)

LINE 10 – A POSITIVE____________________ (Number of Units)

LINE 11 – A NEGATIVE____________________ (Number of Units)

LINE 12 – B POSITIVE ____________________ (Number of Units)

LINE 13 – B NEGATIVE____________________ (Number of Units)

LINE 14 – AB POSITIVE ____________________ (Number of Units)

LINE 15 – AB NEGATIVE ____________________ (Number of Units)

LINE 16 – TOTAL ____________________ (Total Number of Units of Blood Product Being Shipped)

**Repeat lines 7 through 16 as a group when shipping more than one blood product. Assign sequential line numbers to succeeding iterations; for example, first iteration 7 through 16; second iteration 7a through 16a; third iteration 7b through 16b; and so on.

LINE 17 – CONTROL ____________________ (Airbill Number or Transportation Control Number (TCN))

LINE 18 – MISSION ____________________ (Airline and Flight Number or Mission Number Assigned at the Shipment's Origin)

LINE 19 – ARRIVAL____________________ (Estimated DTG of Shipment's Arrival at Destination)

LINE 20 – BOXES ____________________ (Number of Boxes in Shipment)

LINE 21 – CONTACT ____________________ (Name of Shipper's Point of Contact)

LINE 22 – PHONE ____________________ (24-Hour Telephone Number of Shipper's Point of Contact)

LINE 23 – NARRATIVE ____________________ (Free Text for Additional Information Required for Clarification of Report)

LINE 24 – AUTHENTICATION ____________________ (Report Authentication)

TITLE: BULK PETROLEUM ALLOCATION [POLALOT]
REPORT NUMBER: B030 {USMTF # A860}

GENERAL INSTRUCTIONS: Use to allocate bulk petroleum products. Reference: FM 63-20 and FM 63-11.

LINE 1 – DATE AND TIME ________________________________ (DTG)

LINE 2 – UNIT ________________________________ (Unit Making Report)

LINE 3 – COMMAND ________________________________ (Identifier of Component Command)

LINE 4 – FUEL TYPE ________________________________ (Type of Fuel Being Allocated)

LINE 5 – ALLOCATION ________________________________ (Quantity and Unit of Measurement for Fuel Allocated)

LINE 6 – DELIVERY METHOD ________________________________ (Method in Which POL is Delivered)

LINE 7 – LOCATION ________________________________ (UTM or Six-Digit Grid Coordinate With MGRS Grid Zone Designator)

**Repeat lines 3 through 7 to indicate the component command, fuel type, and quantity allocated. Assign sequential line numbers to succeeding iterations; for example, first iteration 3 through 7; second iteration 3a through 7a; third iteration 3b through 7b, and so on.

LINE 8 – NARRATIVE ________________________________ (Free Text for Additional Information Required for Clarification of Report)

LINE 9 – AUTHENTICATION ________________________________ (Report Authentication)

TITLE: BULK PETROLEUM CONTINGENCY REPORT [REPOL]
REPORT NUMBER: B035 {USMTF # B876}

GENERAL INSTRUCTIONS: Use (1) to report loss of major bulk petroleum and (2) to request bulk petroleum products. Reference: FM 63-20 and FM 63-11.

LINE 1 – DATE AND TIME ______________________ (DTG)

LINE 2 – UNIT ______________________ (Unit Making Report)

LINE 3 – NAME ______________________ (Name of Damaged Facility)

LINE 4 – LEVEL ______________________ (Level of Damage)

LINE 5 – REPAIRS ______________________ (Estimated Date Repairs Will be Completed, DTG)

**Repeat lines 3 through 5 to report more than one facility. Assign sequential line numbers to succeeding iterations; for example, first iteration 3 through 5; second iteration 3a through 5a; third iteration 3b through 5b; and so on.

LINE 6 – ASSESSMENT ______________________ (Assessment of Impact of Damage)

LINE 7 – ACTIVITY ______________________ (Name of Activity Reporting Products on Hand)

LINE 8 – PRODUCT ______________________ (Type of Petroleum Product Being Reported)

LINE 9 – INVENT ______________________ (On-Hand Inventory of Product Thousands of Barrels (MBBLs))

LINE 10 – CAPACITY ______________________ (Storage Capacity in MBBLs)

LINE 11 – DAYS ______________________ (Days of Combat Support (DOS) for the Product)

**Repeat lines 7 through 11 to report more than one facility. Assign sequential line numbers to succeeding iterations; for example, first iteration 6 through 10; second iteration 6a through 10a; third iteration 6b through 10b; and so on.

LINE 12 – TYPE ______________________ (Type of Petroleum Product)

LINE 13 – INVENT ______________________ (On-Hand Inventory in Thousands of Barrels (MBBLs))

**Continued on next page.

TITLE: BULK PETROLEUM CONTINGENCY REPORT [REPOL] (CONTINUED)
REPORT NUMBER: B035 {USMTF # B876}

LINE 14 – CAPACITY ____________________ (Usable Storage Capacity in MBBLs)

LINE 15 – DAYS ____________________ (Days of Combat Support (DOS) of Petroleum Products on Hand)

**Repeat lines 12 through 15 to report more than one facility. Assign sequential line numbers to succeeding iterations; for example, first iteration 12 through 15; second iteration 12a through 15a; third iteration, 12b through 15b; and so on.

LINE 16 – DISCHARGE PORT ____________________ (Name of Discharge Port Being Reported)

LINE 17 – PRODUCT ____________________ (Type of Petroleum Product Being Reported)

LINE 18 – BARRELS ____________________ (Quantity of Petroleum Product in Thousands of Barrels (MBBLs))

LINE 19 – PERIOD ____________________ (Period in Which Delivery is Required)

**Repeat lines 16 through 19 to report more than one facility. Assign sequential line numbers to succeeding iterations; for example, first iteration 16 through 19; second iteration 16a through 19a; third iteration 16b through 19b; and so on.

LINE 20 – DAMAGED PORT ____________________ (Name of Damaged Port)

LINE 21 – DAMAGE ____________________ (Level of Damage)

LINE 22 – UNLOAD ____________________ (Unloading Capacity of Damaged Port in Thousands of Barrels (MBBLs))

**Repeat lines 20 through 22 to report more than one facility. Assign sequential line numbers to succeeding iterations; for example, first iteration 20 through 22; second iteration 20a through 22a; third iteration 20b through 22b; and so on.

LINE 23 – NARRATIVE ____________________ (Free Text for Additional Information Required for Clarification of Report)

LINE 24 – AUTHENTICATION ____________________ (Report Authentication)

TITLE: BULK PETROLEUM REQUIREMENTS FORECAST [POLRQMT]
REPORT NUMBER: B040 {USMTF # D869}

GENERAL INSTRUCTIONS: Use to forecast bulk petroleum products. Reference: FM 63-20 and FM 63-11.

LINE 1 – DATE AND TIME ______________________________ (DTG)

LINE 2 – UNIT ______________________________ (Unit Making Report)

LINE 3 – STATUS ______________________________ (Status of Bulk Fuel on Hand to Include Type, Gallons on Hand, Days of Supply on Hand, and Gallons Received Since Last Reported)

TYPE	ON HAND	DAYS	RECEIVED
_______	_______	_______	_______
_______	_______	_______	_______
_______	_______	_______	_______
_______	_______	_______	_______

LINE 4 – FROM ______________________________ (DTG Beginning of Period Covered by Requirement)

LINE 5 – TO ______________________________ (DTG End of Period Covered by Requirement)

LINE 6 – RECEIVING UNIT______________________________ (Designation of the Receiving Unit)

LINE 7 – LOCATION ______________________________ (UTM or Six-Digit Grid Coordinate With MGRS Grid Zone Designator)

**Repeat lines 5 through 7 to indicate other locations, receiving unit designations, and specific requirements for the period shown in line 4. Assign sequential line numbers to succeeding iterations; for example, first iteration 5 through 7; second iteration 5a through 7a; third iteration 5b through 7b, and so on.

**Continued on next page.

TITLE: BULK PETROLEUM REQUIREMENTS FORECAST [POLRQMT] (CONTINUED)
REPORT NUMBER: B040 {USMTF # D869}

LINE 8 – REQUIREMENT ________________________________ (Bulk Fuel Requirement to Include Type, Quantity in Gallons, Transportation Mode Desired, and Priority)

TYPE	QUANTITY	MODE	PRIORITY
______	__________	______	_______
______	__________	______	_______
______	__________	______	_______
______	__________	______	_______

LINE 9 – DAMAGE ____________________________________ (Description of Damage to Bulk Petroleum Facilities and Equipment, to Include Changes in Previous Damage)

LINE 10 – NARRATIVE ________________________________ (Free Text for Additional Information Required for Clarification of Report)

LINE 11 – AUTHENTICATION ___________________________ (Report Authentication)

TITLE: BULK CLASS III REQUEST/FORECAST [BKLIIIREQ]
REPORT NUMBER: B045

GENERAL INSTRUCTIONS: Use to report a periodic update or formal request of fuel unit requirements. Send as needed with ROUTINE precedence on A&L nets. Reference: FM 63-20 and FM 63-11.

LINE 1 – DATE AND TIME ______________________	(DTG)
LINE 2 – UNIT ______________________	(Unit Making Report)
LINE 3 – REQUEST ______________________	(Request or Forecast (Include DTG))
LINE 4 – FUEL TYPE ______________________	(Type Fuel) a) Diesel b) MOGAS c) JP - 4 d) JP - 8
LINE 5 – QTY OH ______________________	(Specify Gallons or Liters)
LINE 6 – QTY REQ ______________________	(Required Quantity)
LINE 7 – NARRATIVE ______________________	(Free Text for Additional Information Required for Clarification of Report)
LINE 8 – AUTHENTICATION ______________________	(Report Authentication)

TITLE: CEMETERY STATUS [CEMSTAT]
REPORT NUMBER: C001 {USMTF # D960}

GENERAL INSTRUCTIONS: Use to provide data concerning the creation or disposition of temporary cemeteries. Reference: FM 10-286 and FM 10-297.

LINE 1 – DATE AND TIME ________________ (DTG)

LINE 2 – UNIT ________________ (Unit Making Report)

LINE 3 – NAME ________________ (Name or Title of Cemetery)

LINE 4 – OPENED ________________ (DTG Cemetery Opened)

LINE 5 – COORDINATES ________________ (UTM or Six-Digit Grid Coordinate With MGRS Grid Zone Designator)

LINE 6 – TOWN NAME ________________ (Name of the Town, Village, or City Nearest the Cemetery)

LINE 7 – ROAD NUMBER ________________ (Cemetery Access Road Number)

LINE 8 – ROAD NAME ________________ (Name of Access Road to Cemetery Site)

LINE 9 – CAPACITY ________________ (Total Capacity of Cemetery)

LINE 10 – REMAINING CAPACITY ________________ (Unused Capacity of the Cemetery)

LINE 11 – DISTANCE ________________ (Distance in Feet Between Grave Reference: Marks)

LINE 12 – WIDTH ________________ (Width in Feet of the Cemetery Access Road)

LINE 13 – SECURITY TYPE ________________ (Security Type Provided at the Cemetery)

**Repeat lines 3 through 13 to report data on multiple temporary cemeteries. Assign sequential line numbers to succeeding iterations; for example, first iteration, 3 through 13; second iteration, 3a through 13a; third iteration, 3b through 13b; and so on.

LINE 14 – CEMETERY NAME ________________ (Name or Title of the Cemetery That Was Closed)

**Continued on next page.

TITLE: CEMETERY STATUS [CEMSTAT] (CONTINUED)
REPORT NUMBER: C001 {USMTF # D960}

LINE 15 – COORDINATES_________________________ (Coordinates of the Cemetery as Closed)

a. _________________________

b. _________________________

c. _________________________

d. _________________________

LINE 16 – DATE CLOSED _________________________ (DTG Cemetery Was Closed)

LINE 17 – QUANTITY BY CATEGORY____________________ (Quantity of Remains by Personnel Category at the Cemetery That Was Closed)

**Repeat lines 14 through 17 to indicate the closing of multiple temporary cemeteries. Assign sequential line numbers to succeeding iterations; for example, first iteration 14 through 17; second iteration 14a through 17a; third iteration 14b through 17b; and so on.

LINE 18 – NAME______________________________ (Name of Cemetery That Was Cleared)

LINE 19 – LOCATION _________________________ (UTM or Six-Digit Grid Coordinate With MGRS Grid Zone Designator of the Cemetery Cleared)

LINE 20 – DATE CLEARED_________________________ (DTG Cemetery Was Cleared)

LINE 21 – LAND CONDITION________________________ (Enter RESTORED or NOT RESTORED)

**Repeat lines 18 through 21 to indicate the clearing of multiple temporary cemeteries. Assign sequential line numbers to succeeding iterations; for example, first iteration 18 through 21; second iteration 18a through 21a; third iteration 18b through 21b; and so on.

LINE 22 – ON HAND____________________________ (Count of Remains on Hand at Beginning of Report Period)

LINE 23 – RECEIVED___________________________ (Count of Remains Received During Period)

**Continued on next page.

TITLE: CEMETERY STATUS [CEMSTAT] (CONTINUED)
REPORT NUMBER: C001 {USMTF # D960}

LINE 24 – TRANSFERRED ______________________________ (Count of Remains Transferred During Period)

LINE 25 – AWAITING TRANSFER__________________________ (Count of Remains Awaiting Transfer at End of Period)

LINE 26 – BURIED ___________________________________ (Count of Remains Buried During Period, if Applicable)

LINE 27 – AWAITING BURIAL____________________________ (Count of Remains Awaiting Burial at End of Period, if Applicable)

**Repeat lines 22 through 27 to indicate multiple reports of remains transferred and buried. Assign sequential line numbers to succeeding iterations; for example, first iteration 22 through 27; second iteration 22a through 27a; third iteration 22b through 27b; and so on.

LINE 28 – NARRATIVE _________________________________ (Free Text for Additional Information Required for Clarification of Report)

LINE 29 – AUTHENTICATION ____________________________ (Report Authentication)

TITLE: CHAPLAIN'S REPORT/UNIT MINISTRY TEAM DAILY LINE REPORT [CHPREP]
REPORT NUMBER: C005

GENERAL INSTRUCTIONS: Use to convey UMT status and general assessment of unit morale by unit chaplain. Send as directed by commander. Reference: FM 16-1.

LINE 1 – DATE AND TIME ______________________________ (DTG)

LINE 2 – UNIT ______________________________ (Unit Making Report)

LINE 3 – UMT STATUS ______________________________ ((Strength) Omit if green)
  a. RED ______________________________ (UMT Loss)
  b. AMBER ______________________________ (Chap or Asst. Loss)
  c. GREEN______________________________ (UMT up, Mission Capable)

LINE 4 – DENOMINATIONAL NEEDS
  a. Protestant ______________________________
  b. Roman Catholic ______________________________
  c. Jewish______________________________
  d. Orthodox ______________________________
  e. Islamic ______________________________
  f. Buddhist______________________________
  g. Other ______________________________ (Specify)

LINE 5 – UNIT RELIGIOUS ACTIVITY NUMBERS
  a. Religious Services
    1. Protestant ______________________________
    2. Roman Catholic ______________________________
    3. Jewish______________________________
    4. Orthodox ______________________________
    5. Islamic ______________________________
    6. Buddhist______________________________
    7. Other ______________________________ (Specify)

  b. Rites, Sacraments, Ordinances
    1. Baptism ______________________________
    2. Communion ______________________________
    3. Last Rites______________________________
    4. Other ______________________________ (Specify)

  c. Pastoral Care
    1. Counseling ______________________________
    2. Casualty Care______________________________
    3. Combat Stress/Battle Fatigue______________________________
    4. Critical Incident Debriefs______________________________
    5. Memorials/Funerals ______________________________
    6. Other ______________________________ (Specify)

**Continued on next page.

TITLE: CHAPLAIN'S REPORT/UNIT MINISTRY TEAM DAILY LINE REPORT [CHPREP] (CONTINUED)
REPORT NUMBER: C005

d. Religious and Humanitarian Support
1. Stability Operations
   a. Civil Affairs ____________________
   b. Noncombatant Evac ____________________
   c. Peace Operations ____________________
   d. Disaster Relief ____________________
   e. Other ____________________ (Specify)
2. Supporting Operations
   a. Humanitarian Assistance ____________________
   b. Environmental Assistance ____________________
   c. Other ____________________ (Specify)

LINE 6 – UMT ASSESSMENT ____________________ (Assessment of Morale, Upcoming Mission, and so forth (if essential))

LINE 7 – NARRATIVE ____________________ (Free Text for Additional Information Required for Clarification of Report)

LINE 8 – AUTHENTICATION ____________________ (Report Authentication)

TITLE: CHEMICAL DOWNWIND REPORT (CDM) [CDMREP]
REPORT NUMBER: C010

GENERAL INSTRUCTIONS: Use to transmit warning of potential chemical hazard from a biological or chemical attack. Send as necessary. Reference: ATP-45(A) and FM 3-3.

LINE 1 – AREA OF VALIDITY ________________________________ (NBC event)

LINE 2 – DATE AND TIME ________________________________ (DTG)

LINE 3 – UNIT ________________________________ (Unit Making Report)

LINE 4 – WINDS ________________________________ (DTG Winds Were Measured)

LINE 5 – FORECAST ________________________________ (DTG Forecast Begins)

LINE 6 – AO ________________________________ (Area of Operation Affected)

LINE 7 – DIRECTION ________________________________ (Downwind Direction)

LINE 8 – WIND SPEED ________________________________ (Wind Speed in K/MPH)

LINE 9 – AIR STABILITY ________________________________ (Air Stability Category:
1 = Very Unstable (U)
2 = Unstable (U)
3 = Slightly Unstable (U)
4 = Neutral (N)
5 = Slightly Stable (S)
6 = Stable (S)
7 = Very Stable (S))

LINE 10 – TEMPERATURE ________________________________ (Temperature Code:

CODE	TEMPERATURE
05	5 DEG C
04	4 DEG C
03	3 DEG C
02	2 DEG C
01	1 DEG C
00	0 DEG C
51	-1 DEG C
52	-2 DEG C
53	-3 DEG C
54	-4 DEG C
55	-5 DEG C
56	-6 DEG C)

**Continued on next page.

TITLE: CHEMICAL DOWNWIND REPORT (CDM) [CDMREP] (CONTINUED)
REPORT NUMBER: C010

LINE 11 – HUMIDITY ____________________	(Humidity Code: 0 = 0-9% 1 = 10-19% 2 = 20-29% 3 = 30-39% 4 = 40-49% 5 = 50-59% 6 = 60-69% 7 = 70-79% 8 = 80-89% 9 = 90-100%)
LINE 12 – WEATHER ____________________	(Significant Weather Phenomena Code: 3 = Blowing Snow and Sand 4 = Fog, Ice Fog, or Thick Haze 5 = Drizzle 6 = Rain 7 = Light Rain or Snow 8 = Showers of Rain, Hail or a Mixture 9 = Thunderstorms (-) = No Significant Phenomena)
LINE 13 – CLOUD COVER ____________________	(Cloud Cover Code)

**Repeat as necessary. Weather information valid for first two hours.

LINE 14 – NARRATIVE ____________________	(Free Text for Additional Information Required for Clarification of Report)
LINE 15 – AUTHENTICATION ____________________	(Report Authentication)

TITLE: CIVIL-MILITARY OPERATIONS STATUS [CIVMILSTAT]
REPORT NUMBER: C015 {USMTF # B991}

GENERAL INSTRUCTIONS: Use to provide a status of the civil affairs and civilian-military cooperation and relations situation. Reference: FM 41-10.

LINE 1 – DATE AND TIME ______	(DTG)
LINE 2 – UNIT ______	(Unit Making Report)
LINE 3 – LIAISON ______	(Establishment of and Changes in Liaison Arrangements Between Civil and Military)
LINE 4 – CONTROL ______	(Establishment and Changes in Location, Status, Activity of Control Nodes and in the CMO AO)
LINE 5 – AREA OF OPERATIONS ______	(CMO AO Name)
LINE 6 – US CIVILIANS ______	(Total Number of US Civilians in AO)
LINE 7 – THIRD COUNTRY ______	(Total Number of Third Country Civilians in AO)
LINE 8 – FOREIGN NATION ______	(Total Number of Foreign Nation Civilians in AO)
LINE 9 – TOTAL CIVILIAN POPULATION ______	(Total Civilian Population in the AO)
LINE 10 – MALES UNDER 16 ______	(Total Number of Dislocated Males Under 16 Years of Age)
LINE 11 – UNACCOMPANIED MALES UNDER 16 ______	(Total Number of Dislocated Unaccompanied Males Under 16 Years of Age)
LINE 12 – MALES OVER 60 ______	(Total Number of Dislocated Males Over 60 Years of Age)
LINE 13 – MARRIED MALES ______	(Total Number of Dislocated Married Males)
LINE 14 – SINGLE MALES ______	(Total Number of Dislocated Single Males)

**Continued on next page.

TITLE: CIVIL-MILITARY OPERATIONS STATUS [CIVMILSTAT] (CONTINUED)
REPORT NUMBER: C015 {USMTF # B991}

LINE 15 – FEMALES UNDER 16 ____________________	(Total Number of Dislocated Females Under 16 Years of Age)
LINE 16 – UNACCOMPANIED FEMALES UNDER 16 __________	(Total Number of Dislocated Unaccompanied Females Under 16 Years of Age)
LINE 17 – FEMALES OVER 60 ____________________	(Total Number of Females Over 60 Years of Age)
LINE 18 – MARRIED FEMALES ____________________	(Total Number of Dislocated Married Females)
LINE 19 – SINGLE FEMALES ____________________	(Total Number of Dislocated Single Females)
LINE 20 – DISLOCATED MALES ____________________	(Total Number of Dislocated Males in AO)
LINE 21 – DISLOCATED FEMALES ____________________	(Total Number of Dislocated Females in AO)
LINE 22 – TOTAL MALES AND FEMALES ______________	(Total Number of Dislocated Males And Females in AO)
LINE 23 – CONTROLLED ____________________	(Total Number of Dislocated Civilians in Controlled Movement)
LINE 24 – UNCONTROLLED ____________________	(Total Number of Dislocated Civilians in Uncontrolled Movement)
LINE 25 – MANPOWER ____________________	(Changes in Civilian Manpower Indicate Changes/Shortages in Labor Availability)
LINE 26 – PUBLIC ADMINISTRATION ______________	(Changes in Capability to Continue Essential Functions)
LINE 27 – LEGAL ____________________	(Changes in Status of Legal Accommodations and Accords)

**Continued on next page.

TITLE: CIVIL-MILITARY OPERATIONS STATUS [CIVMILSTAT] (CONTINUED)
REPORT NUMBER: C015 {USMTF # B991}

LINE 28 – INTELLIGENCE ________________	(Changes in Civilian Attitude; Identification of Collaborators, Agents, and Dissident Elements; Espionage, Sabotage, and Political Subversion)
LINE 29 – LAW AND ORDER ________________	(Changes in Civilian Law and Order)
LINE 30 – PUBLIC HEALTH ________________	(Changes in Status of Public Health)
LINE 31 – SUPPLY ________________	(Changes in Status of Civilian Supplies)
LINE 32 – TRANSPORTATION ________________	(Changes in Status of the Civil Transportation System)
LINE 33 – DRINKING WATER ________________	(Water Potability Status; Potable, Local Only; Nonpotable; to US Standards, as Applicable)
LINE 34 – WATER PRODUCTION ________________	(Adequacy of the Water Production Capability; Adequate, Not Adequate, or None, as Applicable)
LINE 35 – SEWAGE TREATMENT ________________	(Status of Sewage Treatment; Adequate, Not Adequate, or None, as Applicable)
LINE 36 – REFUSE ________________	(Status of Refuse Collection Adequate, Not Adequate, or None as Applicable)
LINE 37 – RADIO ________________	(Status of the Radio Capability Using OPERATIONAL, NONOPERATIONAL, or UNKNOWN)

**Continued on next page.

TITLE: CIVIL-MILITARY OPERATIONS STATUS [CIVMILSTAT] (CONTINUED)
REPORT NUMBER: C015 {USMTF # B991}

LINE 38 – TELEVISION ____________________ (Status of the Television Capability Using OPERATIONAL, NONOPERATIONAL, or UNKNOWN)

LINE 39 – NEWSPAPER ____________________ (Status of the Newspaper Capability Using OPERATIONAL, NONOPERATIONAL, or UNKNOWN)

LINE 40 – POSTAL ____________________ (Status of the Postal Capability Using OPERATIONAL, NONOPERATIONAL, or UNKNOWN)

LINE 41 – TELEPHONE ____________________ (Status of the Telephone Capability Using OPERATIONAL, NONOPERATIONAL, or UNKNOWN)

LINE 42 – TELEGRAPH ____________________ (Status of the Telegraph Capability Using OPERATIONAL, NONOPERATIONAL, or UNKNOWN)

LINE 43 – TELETYPE ____________________ (Status of the Teletype Capability Using OPERATIONAL, NONOPERATIONAL, or UNKNOWN)

LINE 44 – ELECTRIC NAVIGATION SYSTEM ____________ (Status of the Electric Navigation System Using OPERATIONAL, NONOPERATIONAL, or UNKNOWN)

LINE 45 – WORD OF MOUTH ____________________ (Status of Word of Mouth Communications Using OPERATIONAL, NONOPERATIONAL, or UNKNOWN)

**Continued on next page.

TITLE: CIVIL-MILITARY OPERATIONS STATUS [CIVMILSTAT] (CONTINUED)
REPORT NUMBER: C015 {USMTF # B991}

LINE 46 – CIVIL DEFENSE ______________	(Civil Defense Incidents Affecting Civilian Community, Critical Events, and Assistance Provided US Forces)
LINE 47 – SHORTAGES ______________	(Changes or Shortages Community (Water Shortages, Quarantines, Road Restrictions, Curfews))
LINE 48 – SUPPORT RECEIVED______________	(Changes in Support Received From Civilian Community)
LINE 49 – PSYOP ______________	(Changes in Enemy PSYOP Actions Targeted at Civilian Community)
LINE 50 – PROTECTED PROPERTY ______________	(Changes in Status of Arts, Monuments, and Archives)
LINE 51 – CULTURE ______________	(Changes in Local Civil, Social, Cultural, Religious, and Ethnic Characteristics)
LINE 52 – RECOMMENDATIONS ______________	(Narrative Summary and Recommendations Concerning the Status of Civil-Military Operations Activities)
LINE 53 – NARRATIVE ______________	(Free Text for Information Required Clarification of Report)
LINE 54 – AUTHENTICATION ______________	(Report Authentication)

TITLE: CLASS IV BULK BARRIER MATERIALS REQUEST [BLKIVREQ]
REPORT NUMBER: C020

GENERAL INSTRUCTIONS: Use to report a periodic update, forecast, or formal request of Class IV barrier materials a unit requires. Reference: FM 5-102, FM 20-32, and FM 90-7.

LINE 1 – DATE AND TIME ________________ (DTG)

LINE 2 – UNIT ________________ (Unit Making Report)

LINE 3 – REQUEST TYPE ________________ (Update, Forecast, or Request)

LINE 4 – MATERIALS ________________ (Type) a) Pickets
________________ b) Wire
________________ c) Sand Bags
________________ d) Wood Type/Size

LINE 5 – QTY OH ________________ (Specify)

**Repeat as Required

LINE 6 – QTY REQ ________________ (Required Quantity)

**Repeat as Required

LINE 7 – UOM ________________ (Unit of Measurement Times Number Requested)

LINE 8 – MOR ________________ (Method of Resupply Requested (Aerial, LOGPAC))

LINE 9 – LOCATION ________________ (UTM or Six-Digit Grid Coordinate With MGRS Grid Zone Designator to Class IV Supply Point or LRP)

LINE 10 – NLT TIME ________________ (Time Limit For Class IV to Arrive at Supply Point)

LINE 11 – NARRATIVE ________________ (Free Text for Additional Information Required for Clarification of Report)

LINE 12 – AUTHENTICATION ________________ (Report Authentication)

TITLE: CLOSE AIR SUPPORT SUMMARY [CASSUM]
REPORT NUMBER: C025 {USMTF # C600}

GENERAL INSTRUCTIONS: Use to provide timely reports of close, air-support missions and other information obtained during post-flight aircrew debriefing. Reference: FM 6-71.

LINE 1 – DATE AND TIME ______	(DTG)
LINE 2 – UNIT ______	(Unit Making Report)
LINE 3 – NO. CAS MSN ______	(Mission Number for First Mission of the Period and All Remaining Missions)
LINE 4 – NO. AND TYPE ACFT ______	(Number and Type of Aircraft Used)
LINE 5 – SIGHTINGS ______	(Information on Enemy Aircraft and Ground Targets Sighted During Ingress and Egress; Include Location, Direction, Type, Numbers)
LINE 6 – TARGET AREA LOCATION ______	(UTM or Six-Digit Grid Coordinate With MGRS Grid Zone Designator of Target)
LINE 7 – TARGET RESULTS ______	(Type and Quantity of Targets Sighted, Destroyed, or Damaged in the Target Area)
LINE 8 – TARGET DEFENSE ______	(Type, Quantity, and Location of Target Area Defense)
LINE 9 – TARGET WEATHER ______	(Include Low Cloud Cover, Height of Low Cloud Base, Surface Visibility, and the General Weather Condition)
LINE 10 – TARGET AREA JAMMING ______	(Include Type of Jamming Frequencies Jammed, Time Period of Jamming Electronic Protection Action Used, and Effectiveness of Electronic Protection Action Used)

**Continued on next page.

TITLE: CLOSE AIR SUPPORT SUMMARY [CASSUM] (CONTINUED)
REPORT NUMBER: C025 {USMTF # C600}

**Repeat lines 3 through 10 to report more than one CAS mission in a report period. Repeat lines 6 through 10 to report more than one target area within a CAS mission. Assign sequential line numbers to succeeding iterations; for example, first iteration 3 through 10 or 6 through 10; second iteration 3a through 10a or 6a through 10a; and so on.

LINE 11 – ORDNANCE ______________________ (Type and Quantity Expended During Reporting Period)

LINE 12 – NARRATIVE ______________________ (Free Text for Additional Information Required For Clarification of Report)

LINE 13 – AUTHENTICATION ______________________ (Report Authentication)

TITLE: CLOSURE REPORT [CLOSEREP]
REPORT NUMBER: C030

GENERAL INSTRUCTIONS: Use to inform commander of status of movement and capability of the unit to conduct future operations. Submit upon completion of movement as directed by commander. Reference: FM 71-2 and FM 71-3.

LINE 1 – DATE AND TIME ______________	(DTG)
LINE 2 – UNIT ______________	(Unit Making Report)
LINE 3 – SP ______________	(UTM or Six-Digit Grid Coordinate With MGRS Grid Zone Designator to the Start Point and SP Time if Required)
LINE 4 – RP ______________	(UTM or Six-Digit Grid Coordinate With MGRS Grid Zone Designator to the Release Point and RP Time if Required)
LINE 5 – COMMAND POST ______________	(Command Post Location)
LINE 6 – CLOSING DTG ______________	(Closing DTG)
LINE 7 – ACCIDENTS ______________	(Accidents/Incidents/Enemy Activity Encountered)
LINE 8 – ETA TO CONTINUE OPERATIONS ______________	(Estimated Time Unit Will be Reconstituted and Ready to Continue Operations)
LINE 9 – SENSITIVE ITEMS STAT ______________	(Weapons and Sensitive Items Inventory Complete YES/NO/ ANY LOSS)
LINE 10 – NARRATIVE ______________	(Free Text for Additional Information Required for Clarification of Report)
LINE 11 – AUTHENTICATION ______________	(Report Authentication)

TITLE: COMMANDER'S SITUATION REPORT [SITREP]
REPORT NUMBER: C035 {USMTF # C400}

GENERAL INSTRUCTIONS: Use to keep the commander's higher and lower staff updated and advised on the reporting commander's critical situation. Reference: FM 100-15, FM 71-3, and FM 71-2.

LINE 1 – DATE AND TIME ______________________________ (DTG)

LINE 2 – UNIT ______________________________ (Unit Making Report)

LINE 3 – REFERENCE ______________________________ (Provide Reference: Report Title, Originator, and DTG)

LINE 4 – ORIGINATOR ______________________________ (Unit Identification Code of the Unit Originating the Report)

LINE 5 – REPORTED UNIT ______________________________ (Unit Identification Code of the Reported Unit)

LINE 6 – HOME LOCATION ______________________________ (UTM or Six-Digit Grid Coordinate With MGRS Grid Zone Designator for the Home Location of the Reported Unit)

LINE 7 – PRESENT LOCATION ______________________________ (UTM or Six-Digit Grid Coordinate With MGRS Grid Zone Designator for the Present Location of the Reported Unit)

LINE 8 – ACTIVITY ______________________________ (Brief Description of Reported Unit's Current Activity)

LINE 9 – EFFECTIVE ______________________________ (Commander's Evaluation of the Reported Unit's Combat Effectiveness)

**Continued on next page.

TITLE: COMMANDER'S SITUATION REPORT [SITREP] (CONTINUED)
REPORT NUMBER: C035 {USMTF # C400}

LINE 10 – OWN SITUATION DISPOSITION/STATUS ____________	(A Summary Updating Changes to or Not Previously Reported Major Combatant and Support Force Locations; Significant Mission Readiness Degradation on Units; Current Deployments; Proposed Deployments; Changes in Task Force Designations; Organization or Operational Control (CHOP); and Projected Requirements for Additional Forces)
LINE 11 – LOCATION ____________	(UTM or Six-Digit Grid Coordinate With MGRS Grid Zone Designator)
LINE 12 – SITUATION OVERVIEW ____________	(A Brief Overall Assessment of the Situation to Include Circumstances or Conditions Which Increase or Materially Detract From the Capability and Readiness of Forces Assigned or Under Operational Control of the Command or Service)
LINE 13 – OPERATIONS ____________	(A Brief Description and Results of Offensive and Defensive Operations Carried Out by Major Combatant Elements During the Period of the Report; Information on Allied Forces' Operations; Summary of Plans for Combat Operations During Next 24 Hours Including Objectives and Probable Enemy Reaction; Deviations or Variations From Previously Reported Intentions/Plans)

**Continued on next page.

TITLE: COMMANDER'S SITUATION REPORT [SITREP] (CONTINUED)
REPORT NUMBER: C035 {USMTF # C400}

LINE 14 – INTELLIGENCE/RECONNAISSANCE ______________	(Brief Overview of the Situation, Including Operations, Order of Battle, Capabilities, and Threat Changes; Reference: Any Significant Spot Intelligence Reports (SPIREPs) or Intelligence Reports (INTREPs) Submitted in Previous 24 Hours)
LINE 15 – LOGISTICS ______________	(Significant Deficiencies Affecting Support for Planned Operations; Problem Areas Beyond the Commander's or Services' Capability to Overcome or Alleviate in a Timely Manner)
LINE 16 – COMMUNICATIONS/CONNECTIVITY ______________	(Significant Outages, Traffic Volume, Incompatibilities, and Quantitative Equipment Deficiencies; an Assessment of the Mission Impact Caused by Communications Outages and Degradations Should be Provided by the CINC's J-6/J-3 Staff and Contained in This Section)
LINE 17 – PERSONNEL ______________	(Factors Affecting Readiness of Forces/Units; Mobilization Status; Daily Battle Casualties (Ex: KIA, WIA, MIA) Aggregated by Service and Impact of All Casualties Sustained (Battle, Nonbattle, Critical Skills, Key Personnel Upon the Commands' Mission Capability))

**Continued on next page.

TITLE: COMMANDER'S SITUATION REPORT [SITREP] (CONTINUED)
REPORT NUMBER: C035 {USMTF # C400}

LINE 18 – SIGNIFICANT POLITICAL/MILITARY/ DIPLOMATIC EVENTS ______	(Events Not Reported by OPREP 3 PINNACLE But Which Could Result in US and Local and International Public Reaction; Results/Decisions of Key Allied or Other Foreign Government Meetings; Civil Unrest Indications of Civil Defense Measures Contemplated or Implemented; Large-Scale Military Exercises; Events Emphasizing Interests of Key Segments of the Society)
LINE 19 – CDR'S EVAL ______	(Summary of Key Points From Paragraphs 12 Through 19 Highlighting Areas Requiring JCS and NCA Actions or Decisions; Continuity of Operations (COOP) Implementation Intentions on Execution)
LINE 20 – NARRATIVE ______	(Free Text for Information Required for Clarification of Report)
LINE 21 – AUTHENTICATION ______	(Report Authentication)

COMPUTER ATTACK [COMPATK]
REPORT NUMBER: C040

GENERAL INSTRUCTIONS: Use to inform commander and staff of suspected attack into the command information system.

LINE 1 – DATE AND TIME	______________	(DTG)
LINE 2 – UNIT	______________	(Unit Making Report)
LINE 3 – MISSION OF SYSTEM AFFECTED	______________	(Commander's Assessment)
LINE 4 – EQUIPMENT STATUS	______________	(List Type of Equipment Affected)
LINE 5 – DTG ATTACK DETECTED	______________	(DTG Computer System Was Attacked)
LINE 6 – DTG ATTACK ENDED	______________	(DTG Computer Attack Ended)
LINE 7 – SOFTWARE AFFECTED	______________	(List Type and Number of System Software Attacked)
LINE 8 – NUMBER OF SYSTEMS AFFECTED	______________	(List Number of Systems Affected by the Attack)
LINE 9 – TYPE OF ATTACK	______________	(List Type of Attack (Virus Type))
LINE 10 – ADDRESS OF SOURCE	______________	(Address of Attacking Machine)
LINE 11 – INTRUSION METHOD	______________	(Method Used, for Example, Hacked in, From Disk, Internet)
LINE 12 – NARRATIVE	______________	(Free Text for Information Required for Clarification of Report)
LINE 13 – AUTHENTICATION	______________	(Report Authentication)

TITLE: CREW MANNING REPORT [CREWMNQREP]
REPORT NUMBER: C045

GENERAL INSTRUCTIONS: Use to inform commander and staff on status of crews in a unit (such as M1A2, M2A3, and AH-64 crew members). Send as directed by unit and IAW unit SOPs. Reference: FM 17-12-1/2 and FM 23-1.

LINE 1 – DATE AND TIME ______________________________ (DTG)

LINE 2 – UNIT ______________________________ (Unit Making Report)

LINE 3 – CDR'S ASSESSMENT ______________________________ (Commander's Assessment (Bottom Line))

LINE 4 – EQUIPMENT STATUS ______________________________ (Status)

a. EQUIPMENT TYPE (M1A2) ____________________
b. MOS (19K) ____________________
c. PER REQ TO CREW; FULL/MIN (4/3) ____________________
d. NO. CREWS REQUIRED BY TASK ____________________
e. NO. CREWS AVAILABLE; FULL/ MIN ____________________
f. NO. CREWS REQUESTED ____________________

LINE 5 – NARRATIVE ______________________________ (Free Text for Additional Information Required for Clarification of Report)

LINE 6 – AUTHENTICATION ______________________________ (Report Authentication)

TITLE: DAILY BLOOD REPORT [DBLDREP]
REPORT NUMBER: D001

GENERAL INSTRUCTIONS: Use to convey unit blood supply to medical staff and unit commander. Reference: FM 8-10 and FM 8-10-9.

LINE 1 – DATE AND TIME ______________________________ (DTG)

LINE 2 – UNIT ______________________________ (Unit Making Report)

LINE 3 – LOCATION ______________________________ (UTM or Six-Digit Grid Coordinate With MGRS Grid Zone Designator)

LINE 4 – TOTAL NO. ON HAND/ BLOOD PRODUCT/ BLOOD GROUP

______________________________ (Total Number of Blood Products on Hand by Blood Groups at the End of the Reporting Period)

LINE 5 – TOTAL NO. REQUIRED/ BLOOD PRODUCT

______________________________ (Total Number of Blood Products Required by DTG)

LINE 6 – EST TOTAL NO. EXPIRE/ BLOOD PRODUCT/BLOOD GROUP

______________________________ (Estimated Total Number of Blood Products by Blood Groups to Expire Within the Next Seven Days)

LINE 7 – EST TOTAL NO. REQUESTED/ BLOOD PRODUCT/ BLOOD GROUP

______________________________ (Estimated Blood Supply by Quantity and Groups Requested Within the Next Seven Days)

**Repeat lines 4 through 7 to report multiple mission/mission data. Assign sequential line numbers to succeeding iterations; for example, first iteration 4 through 17; second iteration 4a through 7a; third iteration 4b through 7b; and so on.

LINE 8 – CDR'S ASSESSMENT ______________________________ (Commander's Assessment)

LINE 9 – NARRATIVE ______________________________ (Free Text for Additional Information Required for Clarification of Report)

LINE 10 – AUTHENTICATION ______________________________ (Report Authentication)

TITLE: 2406 NMC SUMMARY REPORT [2406NMC]
REPORT NUMBER: D005

GENERAL INSTRUCTIONS: Use to quickly pass critical maintenance information about nonmission capable parts and maintenance. Reference: FM 9-43-1.

LINE 1 – DATE AND TIME ______________________________ (DTG)

LINE 2 – UNIT ______________________________ (Unit Making Report)

LINE 3 – EQUIPMENT TYPE/ AUTH/ON HAND/ NMC/BTL LOSS/ PART NO. STATUS

______________________________ (Equipment Type by Authorized, On Hand, Number Not Mission Capable, Battle Losses, and Part, Number, and Status)

EXAMPLE: M1A1/58/50/3/5/B 65 - 9124-001 P1/C21 - 9125-001 P1/D14 - 9125-017 P1

**Repeat line 3 to report multiple mission/mission data. Assign to succeeding iterations; for example, first iteration 3; second iteration 3a; third iteration 3b; and so on.

LINE 4 – UNIT OR RATE ______________________________ (Unit Operational Readiness (OR) Rate)

LINE 5 – NARRATIVE ______________________________ (Free Text for Information Required for Clarification of Report)

LINE 6 – AUTHENTICATION ______________________________ (Report Authentication)

TITLE: DECONTAMINATION REQUEST [DECONREQ]
REPORT NUMBER: D010 {USMTF # C060}

GENERAL INSTRUCTIONS: Use to request decontamination by contaminated units. Initially request on CMD net, then send all traffic through administrative or logistics net. Send initial request with PRIORITY precedence. Reference: FM 3-5.

LINE 1 – DATE AND TIME ________________________________ (DTG)

LINE 2 – UNIT ________________________________ (Unit Making Report)

LINE 3 – TYPE DECON ________________________________ (Type of Decontamination Support Required (Operational/Thorough))

LINE 4 – TYPE/QUANTITY ________________________________ (Type and Quantity of Equipment to be Decontaminated)

LINE 5 – DECON LOCATION ________________________________ (UTM or Six-Digit Grid Coordinate With MGRS Grid Zone Designator of Decontamination Site Location)

LINE 6 – PERSONNEL ________________________________ (Number of Personnel Requiring Decontamination)

LINE 7 – TYPE CONTAMINATION ________________________________ (Type of Contamination (Nuclear, Biological, or Chemical), if Chemical, Type Agent if Known)

LINE 8 – TIME TO DECON ________________________________ (Request Time to Start Decontaminate)

LINE 9 – LINK-UP POC ________________________________ (Frequency and Call Sign of Link-Up Point POC)

LINE 10 – CP AT DECON SITE ________________________________ (Link-up Point Decontamination Assets)

**Repeat lines 3 through 10 to report multiple mission/mission data. Assign sequential line numbers to succeeding iterations; for example, first iteration 3 through 10; second iteration 3a through 10a; third iteration 3b through 10b; and so on.

LINE 11 – NARRATIVE ________________________________ (Free Text for Additional Information Required for Clarification of Report)

LINE 12 – AUTHENTICATION ________________________________ (Report Authentication)

TITLE: DECON SITE REPORT [DECONSTREP]
REPORT NUMBER: D015 {USMTF # C065}

GENERAL INSTRUCTIONS: Use to update progress and status of decontamination sites in a unit's area of operation. Submit as required IAW unit SOPs. Transmit on O&I ROUTINE precedence. Reference: FM 3-5.

LINE 1 – DATE AND TIME ______________________________ (DTG)

LINE 2 – UNIT ______________________________ (Unit Making Report)

LINE 3 – SUPPORT UNIT ______________________________ (Unit Conducting Decontamination)

LINE 4 – CONTAMINATED UNIT ______________________________ (Unit Being Decontaminated)

LINE 5 – PERCENT COMPLETE ______________________________ (Percentage of Decontamination Complete)

LINE 6 – TIME COMPLETE ______________________________ (DTG of Completion/Site Closure)

LINE 7 – CRITICAL SHORTAGES ______________________________ (Critical Shortages Needed to Complete Decontamination)

LINE 8 – CASUALTIES ______________________________ (Chemical Casualties)

LINE 9 – STATUS ______________________________ (Status of Decontamination Containment (Green, Yellow, Red, Blue))

LINE 10 – TYPE OF DECON ______________________________ (Operational/Thorough)

LINE 11 – NARRATIVE ______________________________ (Free Text for Additional Information Required for Clarification of Report)

LINE 12 – AUTHENTICATION ______________________________ (Report Authentication)

TITLE: DETAINED CIVILIAN PERSONNEL REPORT [DETAINCIVREP]
REPORT NUMBER: D020 {USMTF # C070}

GENERAL INSTRUCTIONS: Use to provide commander of incidents where the unit has detained civilian personnel. Send IAW unit SOPs. Reference: FM 19-40.

LINE 1 – DATE AND TIME	____________________	(DTG)
LINE 2 – UNIT	____________________	(Unit Making Report)
LINE 3 – NO. DETAINEES/ NO. AWAITING RELEASE	____________________	(Number of Personnel Detained and Awaiting Release to Foreign Nation Authorities)
LINE 4 – LOCATION	____________________	(UTM or Six-Digit Grid Coordinate With MGRS Grid Zone Designator of Collection Point)
LINE 5 – REASON FOR DETENTION	____________________	(Reason for Detaining Civilian Personnel)
LINE 6 – MEDICAL CONDITION	____________________	(Report if Detained Persons Require Any Emergency Medical Treatment)
LINE 7 – POC	____________________	(POC of Foreign Nation Support)
LINE 8 – PM ASSESSMENT	____________________	(PM Assessment)
LINE 9 – NARRATIVE	____________________	(Free Text for Additional Information Required for Clarification of Report)
LINE 10 – AUTHENTICATION	____________________	(Report Authentication)

TITLE: DIRECT SUPPORT UNIT REPORT [DERSPTREP]
REPORT NUMBER: D025

GENERAL INSTRUCTIONS: Use by logistics staff officers, combat health support officers, and commanders to report an assessment of the availability and capability of sustaining current and future unit operations. Can be executed as an estimate if required. Each class of supply requires a narrative that includes line number, quantity on-hand, and earliest time on station of resupply of critically short supplies or equipment. Send IAW commander's situation report. Reference: FM 101-5 and FM 700-80.

LINE 1 – DATE AND TIME ________________ (DTG)

LINE 2 – UNIT ________________ (Unit Making Report)

LINE 3 – CDR'S ASSESSMENT ________________ (Commander's Assessment)

LINE 4 – CLASS I ________________ (Class I Narrative)

LINE 5 – CLASS II ________________ (Class II Narrative)

LINE 6 – CLASS III ________________ (Class III Narrative)

LINE 7 – CLASS IV ________________ (Class IV Narrative)

LINE 8 – CLASS V ________________ (Class V Narrative)

LINE 9 – CLASS VI ________________ (Class VI Narrative)

LINE 10 – CLASS VII ________________ (Class VII Narrative)

LINE 11 – CLASS VIII ________________ (Class VIII Narrative)

LINE 12 – CLASS IX ________________ (Class IX Narrative (Include Maintenance Status))

LINE 13 – WATER ________________ (Class I Water)

LINE 14 – TRANSPORTATION ________________ (Transportation Assets Available)

LINE 15 – NARRATIVE ________________ (Free Text for Additional Information Required for Clarification of Report)

LINE 16 – AUTHENTICATION ________________ (Report Authentication)

TITLE: EA (ELECTRONIC ATTACK) DATA MESSAGE [EADAT]
REPORT NUMBER: E001 {USMTF # F751}

GENERAL INSTRUCTIONS: Use to triangulate an object jamming friendly units by comparing lines of bearing from different origins. Reference: FM 34-40-9 and FM 34-54.

LINE 1 – DATE AND TIME ______________________ (DTG)

LINE 2 – UNIT ______________________ (Unit Making Report)

LINE 3 – TRACK ______________________ (Strobe Number)

LINE 4 – BEARING ______________________ (Bearing of the EA Strobe From Affected/Detecting Unit's Position)

LINE 5 – POSITION ______________________ (UTM or Six-Digit Grid Coordinate With MGRS Grid Zone Designator)

LINE 6 – FREQUENCY ______________________ (EA Frequency)

LINE 7 – AFFECTING ______________________ (Type of Equipment Affected by EA, if Known)

LINE 8 – EMITTER ______________________ (Emitter Call Sign and Name or Nomenclature)

LINE 9 – TIME ______________________ (DTG of EA Intercept)

LINE 10 – NARRATIVE ______________________ (Free Text for Additional Information Required for Clarification of Report)

LINE 11 – AUTHENTICATION ______________________ (Report Authentication)

TITLE: EFFECTIVE DOWNWIND MESSAGE [EDM]
REPORT NUMBER: E002 {USMTF # C503}

GENERAL INSTRUCTIONS: Use to provide the command and staff the effective downwind data needed for prediction of fallout areas following NBC attacks. Reference: ATP-45 (A), FM 3-3, and FM 3-5.

LINE 1 – DATE AND TIME ________________________________ (DTG)

LINE 2 – UNIT ________________________________ (Unit Making Report)

LINE 3 – REPORT ________________________________ (Type of Report: EDM, EDF, EDM.NAV, or EDF.NAV)

LINE 4 – AREA ________________________________ (Area of Validity)

LINE 5 – ZULU MIKE ________________________________ (DTG Zulu of Observation and Beginning and End of Period Covered)

LINE 6 – ALFA MIKE ________________________________ (2 KT or Less; Give Radius of Zone 1 in KM/Nautical Miles or Direction and Speed of Wind, With Warning Area Angle When Required)

LINE 7 – BRAVO MIKE ________________________________ (2 KT - 5 KT; Give Radius of Zone 1 in KM/Nautical Miles or Direction and Speed of Wind, With Warning Area Angle When Required)

LINE 8 – CHARLIE MIKE ________________________________ (5 KT - 30 KT; Give Radius of Zone 1 in KM/Nautical Miles or Direction and Speed of Wind, With Warning Area Angle When Required)

LINE 9 – DELTA MIKE ________________________________ (30 KT - 100 KT; Give Radius of Zone 1 in KM/Nautical Miles or Direction and Speed of Wind, With Warning Area Angle When Required)

**Continued on next page.

TITLE: EFFECTIVE DOWNWIND MESSAGE [EDM] (CONTINUED)
REPORT NUMBER: E002 {USMTF # C503}

LINE 10 – ECHO MIKE ________________	(100 KT - 300 KT; Give Radius of Zone 1 in KM/Nautical Miles or Direction and Speed of Wind, With Warning Area Angle When Required)
LINE 11 – FOXTROT MIKE ________________	(300 KT - 1 MT; Give Radius of Zone 1 in KM/Nautical Miles or Direction and Speed of Wind, With Warning Area Angle When Required)
LINE 12 – GOLF MIKE ________________	(1 MT - 3 MT; Give Radius of Zone 1 in KM/Nautical Miles or Direction and Speed of Wind, With Warning Area Angle When Required)
LINE 13 – NARRATIVE ________________	(Free Text for Additional Information Required for Clarification of Report)
LINE 14 – AUTHENTICATION ________________	(Report Authentication)

TITLE: ELECTRONIC WARFARE FREQUENCY DECONFLICTION MESSAGE [EWDECONFLICT]
REPORT NUMBER: E005 {USMTF # F402}

GENERAL INSTRUCTIONS: Use to promulgate a list of protected, guarded, and taboo frequencies to ensure friendly force use of the frequency spectrum without adverse impact from friendly electronic attack. Reference: FM 11-32.

LINE 1 – DATE AND TIME ______________________________ (DTG)

LINE 2 – UNIT ______________________________ (Unit Making Report)

LINE 3 – TYPE ______________________________ (TABOO, PROTECT, or GUARD)

LINE 4 – STATUS ______________________________ (Restricted Status of Frequency: NEW, CHANGE, CANCEL, or RENEW)

LINE 5 – FREQUENCY ______________________________ (Frequency (ies))

LINE 6 – ON TIME ______________________________ (Start DTG of Frequency Restriction)

LINE 7 – OFF TIME ______________________________ (End DTG of Frequency Restriction)

LINE 8 – LOCATION ______________________________ (UTM or Six-Digit Grid Coordinate With MGRS Grid Zone Designator)

**Repeat lines 3 through 8 as a group to accommodate multiple reportings. Assign sequential line numbers to succeeding iterations; for example, first iteration 3 through 8; second iteration 3a through 8a; and so on.

LINE 9 – NARRATIVE ______________________________ (Free Text for Additional Information Required for Clarification of Report)

LINE 10 – AUTHENTICATION ______________________________ (Report Authentication)

TITLE: ELECTRONIC WARFARE MISSION SUMMARY [EWMSNSUM]
REPORT NUMBER: E010 {USMTF # G424}

GENERAL INSTRUCTIONS: Use to summarize significant EW missions and the status of offensive EW assets. Reference: FM 34-10.

LINE 1 – DATE AND TIME ________________ (DTG)

LINE 2 – UNIT ________________ (Unit Making Report)

LINE 3 – FROM ________________ (Beginning DTG Zulu of Period Summarized)

LINE 4 – THROUGH ________________ (Ending DTG Zulu of Period Summarized)

LINE 5 – COUNTRY ________________ (Nationality of the Target Emitter of Concern)

LINE 6 – LOCATION ________________ (UTM or Six-Digit Grid Coordinate With MGRS Grid Zone Designator)

LINE 7 – EMITTER ________________ (Emitter Call Sign and Name or Nomenclature)

LINE 8 – FUNCTION ________________ (Primary Function of Target)

LINE 9 – NOTATION ________________ (Notation or Sorting Code)

LINE 10 – SIGNAL ________________ (Type of Signal of Target Emitter)

LINE 11 – ON TIME ________________ (DTG That Planned EA Activity Was Initiated)

LINE 12 – OFF TIME ________________ (DTG That Planned EA Activity Was Terminated)

LINE 13 – PRIORITY ________________ (Relative Importance of EA Mission)

LINE 14 – TYPE ________________ (Type of EA Used Against the Emitter)

LINE 15 – PRIMARY FREQUENCY ________________ (Primary Frequency of EA Target Signal)

**Continued on next page.

TITLE: ELECTRONIC WARFARE MISSION SUMMARY [EWMSNSUM] (CONTINUED)
REPORT NUMBER: E010 {USMTF # G424}

LINE 16 – SECONDARY FREQUENCY ____________________	(Secondary Frequency of EA Target Signal)
LINE 17 – LOW FREQUENCY ____________________	(Lower Frequency Limit of Target Equipment Class)
LINE 18 – HIGH FREQUENCY ____________________	(Upper Frequency Limit of Target Equipment Class)
LINE 19 – BANDWIDTH ____________________	(Target Frequency Bandwidth Expressed in MHz)
LINE 20 – PULSE REPETITION ____________________	(Pulse Repetition Interval or Frequency)
LINE 21 – SYSTEM USED ____________________	(Name/Nomenclature of EW Asset Used to Perform the Task)
LINE 22 – OPERATIONAL ____________________	(Number of Units That Can Perform Primary EW Mission)
LINE 23 – NONOPERATIONAL ____________________	(Number of Units That Cannot Perform Primary EW Mission)
LINE 24 – DESTROYED ____________________	(Number of Units That Were Destroyed in Combat)
LINE 25 – CHAFF ____________________	(Type of Chaff)
LINE 26 – LOWER FREQUENCY ____________________	(Lower Frequency of a Range of Frequencies That Was Blanked by Chaff or the Lower EA Frequency)
LINE 27 – UPPER FREQUENCY ____________________	(Upper Frequency of a Range of Frequencies That Was Blanked by Chaff or the Lower EA Frequency)
LINE 28 – LOW LEVEL ____________________	(Lower Altitude in Hundreds of Feet of Airspace That Was Blanked by Chaff)
LINE 29 – UPPER LEVEL ____________________	(Upper Altitude in Hundreds of Feet of Airspace That Was Blanked by Chaff)

**Continued on next page.

TITLE: ELECTRONIC WARFARE MISSION SUMMARY [EWMSNSUM] (CONTINUED)
REPORT NUMBER: E010 {USMTF # G424}

LINE 30 – TECHNIQUE ______________________________ (EA Technique Employed)

LINE 31 – COUNTRY ______________________________ (Country in Which Chaff Was Employed)

LINE 32 – ON TIME ______________________________ (DTG That the Chaff Drop Was Initiated)

LINE 33 – OFF TIME______________________________ (DTG That the Chaff Drop Was Terminated)

LINE 34 – START LOCATION______________________________ (Start Location of the Chaff Drop in UTM or Six-Digit Grid Coordinate With MGRS Grid Zone Designator)

LINE 35 – STOP LOCATION ______________________________ (Stop Location of the Chaff Drop in UTM or Six-Digit Grid Coordinate With Grid Zone Designator)

LINE 36 – NARRATIVE ______________________________ (Free Text for Additional Information Required for Clarification of Report)

LINE 37 – AUTHENTICATION ______________________________ (Report Authentication)

TITLE: ELECTRONIC WARFARE REQUESTING/TASKING MESSAGE [EWRTM]
REPORT NUMBER: E015 {USMTF # A426}

GENERAL INSTRUCTIONS: Use (1) to task component commanders to perform EW operations to support the overall EW plan, (2) to support component EW operations, and (3) to request EW support from sources outside their commands. Reference: FM 34-10-2.

LINE 1 – DATE AND TIME ______________________ (DTG)

LINE 2 – UNIT ______________________ (Unit Making Report)

LINE 3 – EA ______________________ (Electronic Activity)

LINE 4 – TASKED ______________________ (Designator of Tasked Unit if the JOC is Tasking the Unit)

LINE 5 – COUNTRY ______________________ (Nationality of the Target Emitter of Concern)

LINE 6 – LOCATION ______________________ (UTM or Six-Digit Grid Coordinate With MGRS Grid Zone Designator)

LINE 7 – EMITTER ______________________ (Emitter Call Sign and Name or Nomenclature)

LINE 8 – FUNCTION ______________________ (Primary Function of Target)

LINE 9 – NOTATION ______________________ (Notation or Sorting Code)

LINE 10 – SIGNAL ______________________ (Type of Signal of Target Emitter)

LINE 11 – ON TIME ______________________ (DTG That Planned EA Activity Was Initiated)

LINE 12 – OFF TIME ______________________ (DTG That Planned EA Activity Will be Terminated)

LINE 13 – PRIORITY ______________________ (Relative Importance of EA Mission)

LINE 14 – TYPE ______________________ (Type of EA and Technique Used Against the Emitter)

LINE 15 – PRIMARY FREQUENCY ______________________ (Primary Frequency of EA Target Signal)

**Continued on next page.

TITLE: ELECTRONIC WARFARE REQUESTING/TASKING REPORT [EWRTM] (CONTINUED)
REPORT NUMBER: E015 {USMTF # A426}

LINE 16 – SECONDARY FREQUENCY ______________________	(Secondary Frequency of EA Target Signal)
LINE 17 – LOW FREQUENCY______________________________	(Lower Frequency Limit of Target Class)
LINE 18 – HIGH FREQUENCY ____________________________	(Upper Frequency Limit of Target Class)
LINE 19 – BANDWIDTH ________________________________	(Target Frequency Bandwidth Expressed in MHz)
LINE 20 – PULSE REPETITION___________________________	(Pulse Repetition Interval or Frequency)
LINE 21 – ES ______________________________________	(Electronic Support)
LINE 22 – COUNTRY _________________________________	(Nationality of the Target Emitter of Concern)
LINE 23 – LOCATION ________________________________	(Emitter Location Described in UTM or Six-Digit Grid Coordinate With MGRS Grid Zone Designator)
LINE 24 – EMITTER __________________________________	(Emitter Call Sign and Name or Nomenclature)
LINE 25 – FUNCTION_________________________________	(Primary Function of Target)
LINE 26 – NOTATION ________________________________	(Notation or Sorting Code)
LINE 27 – SIGNAL __________________________________	(Type of Signal of Target Emitter)
LINE 28 – PRIMARY FREQUENCY ________________________	(Primary Frequency of ES Target Signal)
LINE 29 – SECONDARY FREQUENCY ______________________	(Secondary Frequency of ES Target Signal)
LINE 30 – LOW FREQUENCY_____________________________	(Lower Frequency Limit of Target Equipment Class)

**Continued on next page.

TITLE: ELECTRONIC WARFARE REQUESTING/TASKING REPORT [EWRTM] (CONTINUED)
REPORT NUMBER: E015 {USMTF # A426}

LINE 31 – HIGH FREQUENCY ______________________	(Upper Frequency Limit of Target Class)
LINE 32 – BANDWIDTH ______________________	(Target Frequency Bandwidth Expressed in MHz)
LINE 33 – PULSE REPETITION______________________	(Pulse Repetition Interval or Frequency)
LINE 34 – ON TIME ______________________	(DTG That Planned ES Activity Will be Initiated)
LINE 35 – OFF TIME______________________	(DTG That Planned ES Activity Was Terminated)
LINE 36 – ESSENTIAL ______________________	(EEI Category Indicator)
LINE 37 – PRIORITY ______________________	(Relative Importance of ES Mission)
LINE 38 – CHAFF ______________________	(Type of Chaff)
LINE 39 – LOWER FREQUENCY ______________________	(Lower Frequency of a Range of Frequencies That is to be Blanked by Chaff or the Lower EA Frequency)
LINE 40 – UPPER FREQUENCY ______________________	(Upper Frequency of a Range of Frequencies That is to be Blanked by Chaff or the Upper EA Frequency)
LINE 41 – LOW LEVEL ______________________	(Lower Altitude in Hundreds of Feet of Airspace to be Blanked by Chaff)
LINE 42 – UPPER LEVEL______________________	(Upper Altitude in Hundreds of Feet of Airspace to be Blanked by Chaff)
LINE 43 – TECHNIQUE ______________________	(EA Technique to be Employed)

**Continued on next page.

TITLE: ELECTRONIC WARFARE REQUESTING/TASKING REPORT [EWRTM] (CONTINUED)
REPORT NUMBER: E015 {USMTF # A426}

LINE 44 – COUNTRY ______________________ (Country in Which Chaff is to be Employed)

LINE 45 – ON TIME ______________________ (DTG That the Planned Chaff Drop Will be Initiated)

LINE 46 – OFF TIME______________________ (DTG That the Planned Chaff Drop Will Terminate)

LINE 47 – START LOCATION______________________ (Start Location of the Chaff Drop in UTM or Six-Digit Grid Coordinate With Grid Zone Designator)

LINE 48 – STOP LOCATION ______________________ (Stop Location of the Chaff Drop in UTM or Six-Digit Grid Coordinate With Grid Zone Designator)

LINE 49 – NARRATIVE ______________________ (Free Text for Additional Information Required for Clarification of Report)

LINE 50 – AUTHENTICATION ______________________ (Report Authentication)

TITLE: ENEMY/FRIENDLY/UNIT MINEFIELD/OBSTACLE REPORT [MINOBREP]
REPORT NUMBER: E025

GENERAL INSTRUCTIONS: Use to report all obstacles on the battlefield after developing a report. Disseminate information and report to all command posts and units in the area of operation as soon as possible. Reference: FM 90-13-1, FM 20-32, and FM 5-34.

LINE 1 – DATE AND TIME ____________________ (DTG)

LINE 2 – UNIT ____________________ (Unit Making Report)

LINE 3 – EMPLACING UNIT ____________________ (Emplacing Unit, if Known (Enemy, Friendly Unit, Unit))

LINE 4 – APPROVING AUTHORITY ____________________ (Approving Authority, if Required or Known)

LINE 5 – TARGET/OBSTACLE NO. ____________________ (Target/Obstacle Number, if Required or Known)

LINE 6 – TYPE OF EMPLACING SYSTEM ____________________ (Type of Emplacing System, if Required or Known)

LINE 7 – TYPE MINES/OBSTACLES ____________________ (Type Mines/Obstacle, if Known Include Width and Depth)

LINE 8 – TYPE MARKING SYSTEM ____________________ (Type Minefield/Obstacle Marking System, if Emplaced)

LINE 9 – LIFE CYCLE DTG ____________________ (DTG of Life Cycle/Self-Destruct Time, if Known)

LINE 10 – CORNER LOCATIONS ____________________ (UTM or Six-Digit Grid Coordinate With MGRS Grid Zone Designator of Corners)

LINE 11 – REDUCE ____________________ (Obstacle/Minefield Reduced, YES or NO)

LINE 12 – NO. OF LANES ____________________ (Number of Lanes)

LINE 13 – REDUCTION ASSET USED ____________________ (MICLIC, Mine Plow, Mine Roller, Demolitions, and so on)

LINE 14 – WIDTH ____________________ (Width of Lane)

LINE 15 – DEPTH ____________________ (Depth of Lane)

**Continued on next page.

TITLE: ENEMY/FRIENDLY/UNIT MINEFIELD/OBSTACLE REPORT [MINOBREP] (CONTINUED)
REPORT NUMBER: E025

LINE 16 – GRID TO START OF LANE ______________________ (UTM or Six-Digit Grid Coordinate With MGRS Grid Zone Designator of Start of Lane (Entrance))

LINE 17 – GRID TO END OF LANE ______________________ (UTM or Six-Digit Grid Coordinate With MGRS Grid Zone Designator of End of Lane (Exit))

LINE 18 – LANE MARKING ______________________ (Type of Marking System, if Emplaced)

LINE 19 – BYPASS ______________________ (YES or NO)

LINE 20 – BYPASS GRID ______________________ (UTM or Six-Digit Grid Coordinate With MGRS Grid Zone Designator to Bypass)

LINE 21 – BARRIERS ______________________ (Concertina Wire, Pickets, and/or Trenches, and any other obstacle information necessary)

LINE 22 – NARRATIVE ______________________ (Free Text for Additional Information Required for Clarification of Report)

LINE 23 – AUTHENTICATION ______________________ (Report Authentication)

TITLE: ENEMY/PRISONER OF WAR REPORT [EPOW]
REPORT NUMBER: E030

GENERAL INSTRUCTIONS: Use to convey status and disposition of unit's EPWs. Send IAW unit SOPs. Reference: FM 19-40.

LINE 1 – DATE AND TIME ________________________________ (DTG)

LINE 2 – UNIT ____ (Unit Making Report)

LINE 3 – COUNTRY/ NO. EPW OFF/ __________________________ (By Country of EPW, Number of EPW NCO/ENL/Officers, Number of Lower Enlisted, Total EPW)

a. Country____________________
b. Enemy Officer: GC____/ Field Grade ____/ CO Grade ____/ WO ____/
c. Enemy Enlisted: CSM____/ E-6 – E-8 ____/ E-5 – CPL____/ ENL____/
d. TOTAL NO.___________

LINE 4 – EPW SITE LOCATIONS ____________________________ (EPW UTM or Six-Digit Grid Coordinate With MGRS Grid Zone Designator)

LINE 5 – NO. EPW MISSING ________________________________ (Number EPW Missing or Escaped)

LINE 6 – NO. EPW ESCAPE ATTEMPTS ______________________ (Number EPW Escape Attempts)

LINE 7 - MEDICAL TREATMENT /EVAC _____________________ (Number of EPW Requiring Medical Care and/or Are Transferred to the Combat Health Staff Chain for Medical Evacuation)

LINE 8 – INCIDENT SUMMARY_____________________________ (Summary of Incidents)

LINE 9 – IRC INSPECTION RESULTS ________________________ (International Red Cross Inspections Results, if Required)

**Repeat lines 3 through 9 to report multiple mission/mission data. Assign sequential line numbers to succeeding iterations; for example, first iteration 3 through 9; second iteration 3a through 9a; third iteration 3b through 9b; and so on.

LINE 10 – PM ASSESSMENT_________________________________ (PM Assessment)

**Continued on next page.

TITLE: ENEMY/PRISONER OF WAR REPORT [EPOW] (CONTINUED)
REPORT NUMBER: E030

LINE 11 – NARRATIVE ______________________________ (Free Text for Additional Information Required for Clarification of Report)

LINE 12 – AUTHENTICATION ___________________________ (Report Authentication)

TITLE: ENVIRONMENTAL CONDITION REPORT [ECR]
REPORT NUMBER: E035

GENERAL INSTRUCTIONS: Use to send periodic information (interim snapshots) of the environmental status of specific sites (assembly areas, base camps, logistical support areas, and medical facilities) where hazards are likely to occur, which can result in significant, immediate and/or long-term effects on the natural environment and/or health of friendly forces and noncombatants. Send IAW unit SOPs and commander's direction.

LINE 1 – DATE AND TIME ______________________ (DTG)

LINE 2 – UNIT ______________________ (Unit Making Report)

LINE 3 – LOCATION ______________________ (UTM or Six-Digit Grid Coordinate With MGRS Grid Zone Designator of Site/Incident)

LINE 4 – DESCRIPTION ______________________ (Description of Site/Incident)

LINE 5 – CHANGES ______________________ (Changes From Last ECR or Environmental Baseline Survey EBS)

LINE 6 – HAZARDS ______________________ (Hazards to Natural Environment, Friendly Forces and/or Civilian Personnel)

LINE 7 – ACTIONS ______________________ (Summary of Actions to Minimize Hazards/Remedial Effects)

LINE 8 – UNIT POC ______________________ (Reporting Unit Point of Contact)

LINE 9 – ASSISTANCE ______________________ (Assistance Required/Requested)

LINE 10 – REFERENCE ______________________ (Site-Specific EBS (if required))

LINE 11 – NARRATIVE ______________________ (Free Text for Additional Information Required for Clarification of Report)

LINE 12 – AUTHENTICATION ______________________ (Report Authentication)

TITLE: EXPLOSIVE ORDNANCE DISPOSAL SUPPORT [EODSPT]
REPORT NUMBER: E040 {USMTF # D983}

GENERAL INSTRUCTIONS: Use (1) to request explosive ordinance disposal (EOD) support, (2) to report the results of an EOD mission, or (3) to request EOD support to protect designated very important persons. Reference: FM 9-15.

LINE 1 – DATE AND TIME ______	(DTG)
LINE 2 – UNIT ______	(Unit Making Report)
LINE 3 – ACTIVITY ______	(Type of EOD Activity the Report Concerns: Either EOD REQUEST, EOD RESPONSE, or VIP REQUEST)
LINE 4 – REQUESTOR ______	(Identifier of Unit/Agency Requesting EOD Support)
LINE 5 – EOD UNIT ______	(Identifier of Unit/Agency Performing the EOD Mission)
LINE 6 – CATEGORY ______	(EOD Incident Category Assigned by Requestor; Either INDIRECT, IMMEDIATE, MINOR, or NONE)
LINE 7 – DISCOVERED ______	(DTG Zone When the Unexploded Ordnance Was Discovered)
LINE 8 – DESCRIPTION ______	(If Applicable, Any Additional Descriptive Information Related to the Threat Posed to Resources and Facilities by Unexploded Ordnance)
LINE 9 – ORDNANCE ______	(Number, Type, and Location of Unexploded Ordnance to be Neutralized; Repeat, as Required)
LINE 10 – CONDITION ______	(Either ARMED or UNARMED)

**Continued on next page.

TITLE: EXPLOSIVE ORDNANCE DISPOSAL SUPPORT [EODSPT] (CONTINUED)
REPORT NUMBER: E040 {USMTF # D983}

LINE 11 – SITUATION ________________ (Either DROPPED, IN FIRE, UNDERWATER, ACCIDENT, or a Literal Description of the Circumstances Surrounding the Incident)

**Lines 6 through 11 are applicable if the report is a request for explosive ordnance neutralization.

LINE 12 – REPORTED ________________ (DTG Zone When EOD Incident Was Reported)

LINE 13 – EOD TEAM TIME OF ARRIVAL ________________ (DTG Zone When EOD Team Arrived)

LINE 14 – COMPLETED ________________ (DTG Zone When EOD Action Completed)

LINE 15 – EOD ACTION TAKEN ________________ (Disposition, Condition, Situation, or Other Information Concerning EOD Action Taken)

**Lines 12 through 15 are applicable if the report contains the results of an EOD mission.

LINE 16 – PROTECT ________________ (Title and Last Name of Individual to be Protected)

LINE 17 – NO. OF PERSONNEL ________________ (Number of EOD Personnel Required for Mission)

LINE 18 – DEPART ________________ (Departure Point Name or Coordinates)

LINE 19 – TRANSPORTATION ________________ (Transportation Mode(s) of VIP During Support Period)

LINE 20 – COUNTRIES/AREAS ________________ (Countries/Areas in Which Support is Required)

**Repeat lines 17 through 20 to indicate the number of personnel required, departure point, VIP transportation mode, and country/area requiring support. Assign repetitions in succeeding iterations sequential line numbers; for example, first iteration 17 through 20; second iteration 17a through 20a; third iteration 17b through 20b, and so on.

LINE 21 – BEGIN ________________ (DTG to Begin Zone VIP Support)

**Continued on next page.

TITLE: EXPLOSIVE ORDNANCE DISPOSAL SUPPORT [EODSPT] (CONTINUED)
REPORT NUMBER: E040 {USMTF # D983}

LINE 22 – END____ (DTG to End Zone VIP Support)

**Repeat lines 21 and 22 to indicate the time VIP support is to begin and end. Assign sequential line numbers to succeeding iterations; for example, first iteration 21 through 22; second iteration 21a through 22a; third iteration 21b through 22b, and so on.

LINE 23 – POC____ (Title and Last Name of Individual Designated as the Requesting Unit's POC for Coordinating EOD VIP Support)

LINE 24 – TELEPHONE _________________________________ (Telephone Number of POC)

LINE 25 – FREQUENCY__________________________________ (Primary Radio Frequency of Unit/Agency Requiring EOD Support)

LINE 26 – LOCATION __________________________________ (UTM or Six-Digit Grid Coordinate With MGRS Grid Zone Designator of POC)

**Lines 16 through 26 are applicable if report is a VIP EOD support request.

**Lines 23 through 26 are applicable if a POC has been designated by the unit requesting VIP EOD protection support.

LINE 27 – SPECIAL REQUIREMENTS________________________ (Statement of Any Special Requirements for EOD Support Mission)

LINE 28 – NARRATIVE __________________________________ (Free Text for Additional Information Required for Clarification of Report)

LINE 29 – AUTHENTICATION ____________________________ (Report Authentication)

TITLE: FIRE MISSION-BEACON LOCATION [FM.BEALOC]
REPORT NUMBER: F001 {USMTF # C261}

GENERAL INSTRUCTIONS: Use to report the location and identification code of one or more radar beacons. Reference: FM 6-121.

LINE 1 – DATE AND TIME ________________________________ (DTG)

LINE 2 – UNIT ________________________________ (Unit Making Report)

LINE 3 – BEACON NAME ________________________________ (Beacon Name)

LINE 4 – CODE ________________________________ (Beacon Code)

LINE 5 – LOCATION ________________________________ (Beacon Location, UTM or Six-Digit Grid Coordinate With Grid Zone Designator)

LINE 6 – ON ________________________________ (Relative on Time or Beacon on DTG)

LINE 7 – OFF ________________________________ (Relative off Time or Beacon off DTG)

LINE 8 – NARRATIVE ________________________________ (Free Text for Additional Information Required for Clarification of Report)

**Repeat lines 3 through 8 to report information on multiple beacons. Assign sequential line numbers to succeeding iterations; for example, first iteration 3 through 8; second iteration 3a through 8a, and so on.

LINE 9 – AUTHENTICATION ________________________________ (Report Authentication)

TITLE: FIRE MISSION-REQUEST TO FIRE [FM.RF]
REPORT NUMBER: F005 {USMTF # D212}

GENERAL INSTRUCTIONS: Use to request permission to fire across common boundaries from an adjacent service unit. Reference: FM 6-30.

LINE 1 – DATE AND TIME ______________________ (DTG)

LINE 2 – UNIT ______________________ (Unit Making Report)

LINE 3 – TARGET NO. ______________________ (Fire Support Target Number)

LINE 4 – GRID ______________________ (UTM or Six-Digit Grid Coordinate With MGRS Grid Zone Designator)

LINE 5 – DESCRIPTION ______________________ (Target Description)

LINE 6 – FIRING UNIT ______________________ (Firing Unit)

LINE 7 – WEAPON ______________________ (Weapon Type)

LINE 8 – ROUNDS IN EFFECT ______________________ (Number of Rounds in Effect)

LINE 9 – PROJECTILE ______________________ (Projectile Type)

LINE 10 – FUZE ______________________ (Fuze Type)

LINE 11 – EST DTG INITIAL RDS ______________________ (Estimated DTG of Initial Rounds)

LINE 12 – NARRATIVE ______________________ (Free Text for Additional Information Required for Clarification of Report)

LINE 13 – AUTHENTICATION ______________________ (Report Authentication)

TITLE: FIRE PLANNING-COMPUTE A FIRE PLAN [FP.COMPFP]
REPORT NUMBER: F010 {USMTF # A272}

GENERAL INSTRUCTIONS: Use to specify the h-hour, the number of fire plan phases, the phases start times relative to h-hour, and the phase lengths. Reference: FM 6-20 and FM 6-20-10.

LINE 1 – DATE AND TIME	________________	(DTG)
LINE 2 – UNIT	________________	(Unit Making Report)
LINE 3 – POI	________________	(Primary Option Indicator. Enter ADD, AMEND, or CANCEL)
LINE 4 – PLAN	________________	(Fire Plan Designation)
LINE 5 – H-HOUR	________________	(H-Hour Fire Plan Operation Order)
LINE 6 – ALFA START	________________	(Phase Start Time)
LINE 7 – LENGTH	________________	(Fire Plan Phase Length in Minutes)
LINE 8 – BRAVO START	________________	(Phase Start Time)
LINE 9 – LENGTH	________________	(Fire Plan Phase Length in Minutes)
LINE 10 – CHARLIE START	________________	(Phase Start Time)
LINE 11 – LENGTH	________________	(Fire Plan Phase Length in Minutes)
LINE 12 – DELTA START	________________	(Phase Start Time)
LINE 13 – LENGTH	________________	(Fire Plan Phase Length in Minutes)
LINE 14 – NARRATIVE	________________	(Free Text for Additional Information Required for Clarification of Report)
LINE 15 – AUTHENTICATION	________________	(Report Authentication)

TITLE: FIRE PLANNING-FIRE PLAN EXECUTIVE ORDERS [FP.FPO]
REPORT NUMBER: F015 {USMTF # E272}

GENERAL INSTRUCTIONS: Use (1) to identify the firing element and (2) to specify the number of rounds or effects required for each target in a fire plan. Reference: FM 6-20 and FM 6-20-10.

LINE 1 – DATE AND TIME ______________________________ (DTG)

LINE 2 – UNIT ______________________________ (Unit Making Report)

LINE 3 – POI ______________________________ (Primary Option Indicator. Enter ADD, AMEND, or CANCEL)

LINE 4 – PLAN ______________________________ (Fire Plan Designation)

LINE 5 – UNIT DESIGNATORS______________________________ (Section, Platoon, Battery/Company, Battalion/Regiment and Regiment/Brigade/Division Designators, or Ship Call Sign(s))

LINE 6 – UNIT DESIGNATORS______________________________ (Section, Platoon, Battery/Company, Battalion/Regiment and Regiment/Brigade/Division Designators, or Ship Call Sign(s))

LINE 7 – UNIT DESIGNATORS______________________________ (Section, Platoon, Battery/Company, Battalion/Regiment and Regiment/Brigade/Division Designators, or Ship Call Sign(s))

LINE 8 – UNIT DESIGNATORS______________________________ (Section, Platoon, Battery/Company, Battalion/Regiment and Regiment/Brigade/Division Designators, or Ship Call Sign(s))

LINE 9 – UNIT DESIGNATORS______________________________ (Section, Platoon, Battery/Company, Battalion/Regiment and Regiment/Brigade/Division Designators, or Ship Call Sign(s))

LINE 10 – TARGET NO. ______________________________ (Fire Support Target Number)

**Continued on next page.

TITLE: FIRE PLANNING-FIRE PLAN EXECUTIVE ORDERS [FP.FPO] (CONTINUED)
REPORT NUMBER: F015 {USMTF # E272}

LINE 11 – TARGET ID ______ (Target Identification Number)

LINE 12 – TRAJECTORY ______ (Type of Trajectory)

LINE 13 – TRAJECTORY ______ (Type of Trajectory)

LINE 14 – TRAJECTORY ______ (Type of Trajectory)

LINE 15 – TRAJECTORY ______ (Type of Trajectory)

LINE 16 – TRAJECTORY ______ (Type of Trajectory)

LINE 17 – VOLLEYS ______ (Number of Initial Volleys and Number of Subsequent Volleys)

LINE 18 – VOLLEYS ______ (Number of Initial Volleys and Number of Subsequent Volleys)

LINE 19 – VOLLEYS ______ (Number of Initial Volleys and Number of Subsequent Volleys)

LINE 20 – VOLLEYS ______ (Number of Initial Volleys and Number of Subsequent Volleys)

LINE 21 – VOLLEYS ______ (Number of Initial Volleys and Number of Subsequent Volleys)

LINE 22 – EFFECTS ______ (Effects (Percent Damage) Required)

LINE 23 – EFFECTS ______ (Effects (Percent Damage) Required)

LINE 24 – EFFECTS ______ (Effects (Percent Damage) Required)

LINE 25 – EFFECTS ______ (Effects (Percent Damage) Required)

LINE 26 – EFFECTS ______ (Effects (Percent Damage) Required)

LINE 27 – RELATIVE TIME ______ (Time in Minutes Relative to H-Hour)

**Continued on next page.

TITLE: FIRE PLANNING-FIRE PLAN EXECUTIVE ORDERS [FP.FPO] (CONTINUED)
REPORT NUMBER: F015 {USMTF # E272}

LINE 28 – PHASE ______________________________ (Phase of Fire)

LINE 29 – SHELL ______________________________ (Initial Volley Projectile and Subsequent Volley Projectile)

LINE 30 – SHELL ______________________________ (Initial Volley Projectile and Subsequent Volley Projectile)

LINE 31 – SHELL ______________________________ (Initial Volley Projectile and Subsequent Volley Projectile)

LINE 32 – SHELL ______________________________ (Initial Volley Projectile and Subsequent Volley Projectile)

LINE 33 – SHELL ______________________________ (Initial Volley Projectile and Subsequent Volley Projectile)

LINE 34 – FUZE______________________________ (Initial Volley Fuze and Subsequent Volley Fuze)

LINE 35 – FUZE______________________________ (Initial Volley Fuze and Subsequent Volley Fuze)

LINE 36 – FUZE______________________________ (Initial Volley Fuze and Subsequent Volley Fuze)

LINE 37 – FUZE______________________________ (Initial Volley Fuze and Subsequent Volley Fuze)

LINE 38 – FUZE______________________________ (Initial Volley Fuze and Subsequent Volley Fuze)

LINE 39 – TARGET LIST ______________________________ (Target List/Last Target Indicator)

**Repeat lines 10 through 39 if necessary. Assign sequential line numbers to succeeding iterations; for example, first iteration 10 through 39; second iteration 10a through 39a; and so on.

LINE 40 – NARRATIVE ______________________________ (Free Text for Additional Information Required for Clarification of Report)

LINE 41 – AUTHENTICATION ______________________________ (Report Authentication)

TITLE: FIRE PLANNING-FIRE PLAN TARGET LIST [FP.FPT]
REPORT NUMBER: F020 {USMTF # E271}

GENERAL INSTRUCTIONS: Use (1) to list targets in the Fire Plan Target List and (2) to give scheduling data of fires in a fire plan. Reference: FM 6-20 and FM 6-20-10.

LINE 1 – DATE AND TIME ________________ (DTG)

LINE 2 – UNIT ________________ (Unit Making Report)

LINE 3 – POI ________________ (Primary Option Indicator. Enter ADD, AMEND, or CANCEL)

LINE 4 – PLAN ________________ (Fire Plan Designation)

LINE 5 – TARGET NO. ________________ (Fire Support Target Number)

LINE 6 – TARGET ID ________________ (Target Identification Number)

LINE 7 – ZONE ________________ (UTM Grid Zone and 100-KM Square)

LINE 8 – GRID ________________ (UTM or Six-Digit Grid Coordinate With MGRS Grid Zone Designator)

LINE 9 – TARGET ELEMENTS ________________ (Number of Target Elements)

LINE 10 – TYPE ________________ (Target or Friendly Unit Type and Subtype)

LINE 11 – PROTECTION ________________ (Degree of Personnel Protection)

LINE 12 – RADIUS ________________ (Target Radius in Meters)

LINE 13 – LENGTH ________________ (Target Length in Meters)

LINE 14 – WIDTH ________________ (Target Width in Meters)

LINE 15 – ATTITUDE ________________ (Target Attitude in Mils)

LINE 16 – ACCURACY ________________ (Target Report Accuracy in Meters)

LINE 17 – RELATIVE TIME ________________ (Time in Minutes Relative to H-Hour)

**Continued on next page.

TITLE: FIRE PLANNING-FIRE PLAN TARGET LIST [FP.FPT] (CONTINUED)
REPORT NUMBER: F020 {USMTF # E271}

LINE 18 – PRIORITY ________________ (Fire Plan Target Priority)

LINE 19 – PHASE ________________ (Phase of Fire)

LINE 20 – PHASE ________________ (Phase of Fire)

LINE 21 – PHASE ________________ (Phase of Fire)

LINE 22 – PHASE ________________ (Phase of Fire)

LINE 23 – GROUP ________________ (Target Group Number)

LINE 24 – SERIES ________________ (Target Series Name and Attack Sequence Number)

LINE 25 – POSITION ________________ (Single Firing Site Designator)

LINE 26 – TARGET LIST ________________ (Target List/Last Target Indicator)

**Repeat lines 5 through 26 if necessary. Assign sequential line numbers to succeeding iterations; for example, first iteration 5 through 26; second iteration 5a through 26a; and so on.

LINE 27 – NARRATIVE ________________ (Free Text for Additional Information Required for Clarification of Report)

LINE 28 – AUTHENTICATION ________________ (Report Authentication)

TITLE: FIRE PLANNING-NUCLEAR SCHEDULE |FP.NUCSCD|
REPORT NUMBER: F025 {USMTF # D271}

GENERAL INSTRUCTIONS: Use to add, amend, or cancel the specified nuclear target in a specified fire plan. Reference: FM 6-20.

LINE 1 – DATE AND TIME ________________ (DTG)

LINE 2 – UNIT ________________ (Unit Making Report)

LINE 3 – POI ________________ (Primary Option Indicator. Enter ADD, AMEND, or CANCEL)

LINE 4 – PLAN ________________ (Fire Plan Designation)

LINE 5 – TARGET NO. ________________ (Fire Support Target Number)

LINE 6 – ZONE ________________ (UTM Grid Zone and 100-KM Square)

LINE 7 – GRID ________________ (UTM or Six-Digit Grid Coordinate With MGRS Grid Zone Designator)

LINE 8 – ALTITUDE ________________ (Altitude in Meters)

LINE 9 – DGZ GZ & 100-KM SQ ________________ (Desired Ground Zero - UTM Grid Zone and 100-KM Square)

LINE 10 – DGZ 1M E & 1M N ________________ (Desired Ground Zero - UTM 1-Meter Easting and UTM 1-Meter Northing)

LINE 11 – GZ ALTITUDE ________________ (Ground Zero Altitude in Meters)

LINE 12 – TARGET ELEMENTS ________________ (Number of Target Elements)

LINE 13 – TYPE ________________ (Target or Friendly Unit Type and Subtype)

LINE 14 – PROTECTION ________________ (Degree of Personnel Protection)

LINE 15 – RADIUS ________________ (Target or Friendly Unit Radius)

**Continued on next page.

TITLE: FIRE PLANNING-NUCLEAR SCHEDULE [FP.NUCSCD] (CONTINUED)
REPORT NUMBER: F025 {USMTF # D271}

LINE 16 – LENGTH__________ (Target or Friendly Unit Length)

LINE 17 – WIDTH __________ (Target or Friendly Unit Width)

LINE 18 – ATTITUDE__________ (Attitude in Mils)

LINE 19 – TIME__________ (Time in Minutes Relative to H-Hour)

LINE 20 – DURATION __________ (Duration of Fire)

LINE 21 – UNIT DESIGNATORS__________ (Section, Platoon, Battery/ Company, Battalion/Regiment and Regiment/Brigade/Division Designators)

LINE 22 – POSITION__________ (UTM or Six-Digit Grid Coordinate With Grid Zone Single Firing Site Designator)

LINE 23 – FIRE FOR EFFECT __________ (Fire for Effect Number of Volleys)

LINE 24 – PROJECTILE __________ (Fire for Effect Projectile)

LINE 25 – FUZE__________ (Fire for Effect Fuze)

LINE 26 – NARRATIVE __________ (Free Text for Additional Information Required for Clarification of Report)

LINE 27 – AUTHENTICATION __________ (Report Authentication)

TITLE: FIRE PLANNING-RESERVE FIRE UNIT [FP.RESFU]
REPORT NUMBER: F030 {USMTF # A271}

GENERAL INSTRUCTIONS: Use to preclude use of fire units during a specific time interval in a specific fire plan. Reference: FM 6-20.

LINE 1 – DATE AND TIME ______________________ (DTG)

LINE 2 – UNIT ______________________ (Unit Making Report)

LINE 3 – POI ______________________ (Primary Option Indicator. Enter ADD, AMEND, or CANCEL)

LINE 4 – PLAN ______________________ (Fire Plan Designation)

LINE 5 – UNIT DESIGNATORS______________________ (Section, Platoon, Battery/ Company, Battalion/Regiment and Regiment/Brigade/Division Designators, or Ship Call Sign(s))

LINE 6 – START ______________________ (Reservation Start Time in Minutes)

LINE 7 – STOP______________________ (Reservation Stop Time in Minutes)

**Repeat lines 5 through 7 to provide reservation information on multiple fire units. Assign sequential line numbers to succeeding iterations; for example, first iteration 5 through 7; second iteration 5a through 7a; third iteration 5b through 7b, and so on.

LINE 8 – NARRATIVE ______________________ (Free Text for Additional Information Required for Clarification of Report)

LINE 9 – AUTHENTICATION ______________________ (Report Authentication)

TITLE: FIRE SUPPORT ELEMENT-COMMANDER'S CRITERIA [FSE.CRITER]
REPORT NUMBER: F035 {USMTF # E273}

GENERAL INSTRUCTIONS: Use to disseminate analysis criteria for nuclear target analysis, chemical target analysis, nuclear fire planning, and fallout prediction. Reference: FM 6-20.

LINE 1 – DATE AND TIME	______________________	(DTG)
LINE 2 – UNIT	______________________	(Unit Making Report)
LINE 3 – POI	______________________	(Primary Option Indicator. Enter ADD, AMEND, or CANCEL)
LINE 4 – PLAN	______________________	(Fire Plan Designation)
LINE 5 – TIME	______________________	(DTG of Fire Plan)
LINE 6 – TARGET NO.	______________________	(Fire Support Target Number)
LINE 7 – UNIT DESIGNATORS	______________________	(Section, Platoon, Battery/ Company, Battalion/Regiment and Regiment/Brigade/Division Designators)
LINE 8 – CIRCULAR ERROR PROBABLE	______________________	(Nuclear Target Analysis Circular Error Probable)
LINE 9 – ZONE	______________________	(Zone of Fire Responsibility)
LINE 10 – EXCEPTION	______________________	(Nuclear Target Analysis Exception Indicator)
LINE 11 – EXCEPTION	______________________	(Nuclear Target Analysis Exception Indicator)
LINE 12 – EXCEPTION	______________________	(Nuclear Target Analysis Exception Indicator)
LINE 13 – EXCEPTION	______________________	(Nuclear Target Analysis Exception Indicator)
LINE 14 – AREA	______________________	(Required Coverage for Area Targets)
LINE 15 – CRITERIA	______________________	(Personnel Radiation Casualty Criteria)

**Continued on next page.

TITLE: FIRE SUPPORT ELEMENT-COMMANDER'S CRITERIA [FSE.CRITER] (CONTINUED)
REPORT NUMBER: F035 {USMTF # E273}

LINE 16 – PRIMARY ______	(Primary Target Category)
LINE 17 – PROBABILITY ______	(Required Probability for Point Targets)
LINE 18 – DAMAGE ______	(Material Damage Level)
LINE 19 – VULNERABILITY ______	(Troop Safety Vulnerability Category)
LINE 20 – RISK ______	(Troop Safety Risk Category)
LINE 21 – VEGETATION ______	(Vegetation Flammability Category Designator and Forest Type)
LINE 22 – MAXIMUM YIELD ______	(Maximum Yield to be Considered)
LINE 23 – MINIMUM DISTANCE ______	(Minimum Distance Allowable for Scheduling (Adjusted))
LINE 24 – PRECIPITATION ______	(Precipitation Indicator)
LINE 25 – WEAPON ______	(Surface-to-Surface Weapon Type)
LINE 26 – MAX VOLLEYS ______	(Maximum Number of Volleys per Fire Unit)
LINE 27 – CHEMICAL MINIMUM ______	(Minimum Allowable Chemical Effects)
LINE 28 – OPTIMUM ______	(Optimum Chemical Effects)
LINE 29 – MAX CHEM ______	(Maximum Allowable Chemical Effects and Persistent Indicator)
LINE 30 – NARRATIVE ______	(Free Text for Additional Information Required for Clarification of Report)
LINE 31 – AUTHENTICATION ______	(Report Authentication)

TITLE: FIRE SUPPORT ELEMENT-FRIENDLY UNIT LOCATION [FSE.FRD]
REPORT NUMBER: F040 {USMTF # B260}

GENERAL INSTRUCTIONS: Use to report information concerning friendly unit locations and status (for other than organic artillery and NSF units) to facilitate NBC vulnerability analysis. Reference: FM 6-20.

LINE 1 – DATE AND TIME ________________________________ (DTG)

LINE 2 – UNIT ________________________________ (Unit Making Report)

LINE 3 – POI ________________________________ (Primary Option Indicator. Enter ADD, AMEND, or CANCEL)

LINE 4 – UNIT DESIGNATORS ________________________________ (Section, Platoon, Battery/ Company, Battalion/Regiment and Regiment/Brigade/Division Designators)

LINE 5 – AIR UNIT DESIGNATORS ________________________________ (Friendly Unit Designator – Used Only for Air Units)

LINE 6 – SHIP CALL SIGN ________________________________ (Fire Support Ship Call Sign(s))

LINE 7 – PLAN ________________________________ (Fire Plan Designation)

LINE 8 – ZONE ________________________________ (UTM Grid Zone and 100-KM Square)

LINE 9 – GRID ________________________________ (UTM or Six-Digit Grid Coordinate With MGRS Grid Zone Designator)

OR (IF NECESSARY)

LINE 10 – LAT/LONG ________________________________ (Unit Location, LAT/LONG)

LINE 11 – TYPE/SUBTYPE ________________________________ (Friendly Unit Type, Friendly Unit Subtype)

LINE 12 – PROTECTION ________________________________ (Degree of Personnel Protection)

LINE 13 – RADIUS ________________________________ (Friendly Unit Radius in Meters)

LINE 14 – LENGTH ________________________________ (Friendly Unit Length in Meters)

**Continued on next page.

TITLE: FIRE SUPPORT ELEMENT-FRIENDLY UNIT LOCATION [FSE.FRD] (CONTINUED)
REPORT NUMBER: F040 {USMTF # B260}

LINE 15 – WIDTH ________________ (Friendly Unit Width in Meters)

LINE 16 – ATTITUDE ________________ (Target Attitude in Mils)

LINE 17 – SAFETY VULNERABILITY ________________ (Troop Safety Vulnerability Category)

LINE 18 – SAFETY RISK ________________ (Troop Safety Risk Category)

LINE 19 – RADIATION STATUS ________________ (Nuclear Radiation Exposure Status)

LINE 20 – NUC VULNERABILITY ________________ (Nuclear Vulnerability Category)

LINE 21 – NARRATIVE ________________ (Free Text for Additional Information Required for Clarification of Report)

LINE 22 – AUTHENTICATION ________________ (Report Authentication)

TITLE: FLIGHT CONTROL INFORMATION [FLTCONTINFO]
REPORT NUMBER: F045 {USMTF # F632}

GENERAL INSTRUCTIONS: Use to provide control and coordination information for aircraft entering another service or component's airspace. Reference: FM 100-103-1.

LINE 1 – DATE AND TIME ____________________ (DTG)

LINE 2 – UNIT ____________________ (Unit Making Report)

LINE 3 – PLAN ____________________ (Flight Plan Designation)

LINE 4 – MISSION ____________________ (Mission Number)

LINE 5 – STATUS ____________________ (Status of Flight Plan: APPROVED Or DISAPPROVED)

LINE 6 – CONTROL ____________________ (Type Control: INITIAL, FINAL, COORDINATION, MONITOR)

LINE 7 – CALL SIGN ____________________ (Call Sign of Control Agency)

LINE 8 – PRIMARY____________________ (Primary Frequency or Frequency Designator)

LINE 9 – SECONDARY____________________ (Secondary Frequency or Frequency Designator)

LINE 10 – POSITION____________________ (UTM or Six-Digit Grid Coordinate With Grid Zone Designator at Which the Aircraft Should Contact the Control Agency)

LINE 11 – CONTROL ____________________ (Type Control: INITIAL, FINAL, COORDINATION, MONITOR)

LINE 12 – CALL SIGN ____________________ (Call Sign of Control Agency)

LINE 13 – PRIMARY____________________ (Primary Frequency or Frequency Designator)

LINE 14 – SECONDARY____________________ (Secondary Frequency or Frequency Designator)

**Continued on next page.

TITLE: FLIGHT CONTROL INFORMATION [FLTCONTINFO] (CONTINUED)
REPORT NUMBER: F045 {USMTF # F632}

LINE 15 – POSITION________________________	(UTM or Six-Digit Grid Coordinate With MGRS Grid Zone Designator at Which the Aircraft Should Contact the Control Agency)
LINE 16 – CONTROL ________________________	(Type Control: INITIAL, FINAL, COORDINATION, MONITOR)
LINE 17 – CALL SIGN ________________________	(Call Sign of Control Agency)
LINE 18 – PRIMARY________________________	(Primary Frequency or Frequency Designator)
LINE 19 – SECONDARY________________________	(Secondary Frequency or Frequency Designator)
LINE 20 – POSITION________________________	(UTM or Six-Digit Grid Coordinate With MGRS Grid Zone Designator at Which the Aircraft Should Contact the Control Agency)
LINE 21 – CONTROL ________________________	(Type Control: INITIAL, FINAL, COORDINATION, MONITOR)
LINE 22 – CALL SIGN ________________________	(Call Sign of Control Agency)
LINE 23 – PRIMARY________________________	(Primary Frequency or Frequency Designator)
LINE 24 – SECONDARY________________________	(Secondary Frequency or Frequency Designator)
LINE 25 – POSITION________________________	(UTM or Six-Digit Grid Coordinate With MGRS Grid Zone Designator at Which the Aircraft Should Contact the Control Agency)
LINE 26 – CONTROL ________________________	(Type Control: INITIAL, FINAL, COORDINATION, MONITOR)

**Continued on next page.

TITLE: FLIGHT CONTROL INFORMATION [FLTCONTINFO] (CONTINUED)
REPORT NUMBER: F045 {USMTF # F632}

LINE 27 – CALL SIGN ______	(Call Sign of Control Agency)
LINE 28 – PRIMARY ______	(Primary Frequency or Frequency Designator)
LINE 29 – SECONDARY ______	(Secondary Frequency or Frequency Designator)
LINE 30 – POSITION ______	(UTM or Six-Digit Grid Coordinate With MGRS Grid Zone Designator at Which the Aircraft Should Contact the Control Agency)
LINE 31 – CONTROL ______	(Type Control: INITIAL, FINAL, COORDINATION, MONITOR)
LINE 32 – CALL SIGN ______	(Call Sign of Control Agency)
LINE 33 – PRIMARY ______	(Primary Frequency or Frequency Designator)
LINE 34 – SECONDARY ______	(Secondary Frequency or Frequency Designator)
LINE 35 – POSITION ______	(UTM or Six-Digit Grid Coordinate With MGRS Grid Zone Designator at Which the Aircraft Should Contact the Control Agency)
LINE 36 – NARRATIVE ______	(Free Text for Additional Information Required for Clarification of Report)
LINE 37 – AUTHENTICATION ______	(Report Authentication)

TITLE: FRAGMENTARY ORDER [FRAGO]
REPORT NUMBER: F050

GENERAL INSTRUCTIONS: Use to send timely changes of existing orders to subordinate and supporting commanders while providing notification to higher and adjacent commands. Send FLASH precedence O&I or CMD IAW unit SOPs. Reference: FM 101-5.

LINE 1 – DATE AND TIME ________________ (DTG)

LINE 2 – UNIT ________________ (Unit Making Report)

LINE 3 – FRAGO ORDER NO. ________________ (Fragmentary Order)

LINE 4 – REFERENCES ________________ (Changes to OPORD Only)

LINE 5 – TIME ZONE ________________ (Time Zone Used in FRAGO)

LINE 6 – SITUATION ________________ (Mandatory Include Changes)

LINE 7 – MISSION ________________ (Mandatory)

LINE 8 – EXECUTION – INTENT ________________ (Optional)

a. CONCEPT OF OPERATIONS ________________ (Changes to OPORD Only)

b. TASKS TO SUB UNITS ________________ (Changes to OPORD Only)

c. COORDINATING INSTRUCTIONS. ________________ (Changes to OPORD Only)

LINE 9 – SERVICE SUPPORT ________________ (With Changes)

LINE 10 – COMMAND AND SIGNAL ________________ (With Changes)

LINE 11 – ACKNOWLEDGE ________________ (Mandatory)

LINE 12 – CDR NAME, RANK ________________ (Commander's Name and Rank)

LINE 13 – OFFICIAL ________________ (Optional)

LINE 14 – ANNEXES ________________ (Optional)

LINE 15 – DISTRIBUTION ________________ (Optional)

LINE 16 – NARRATIVE ________________ (Free Text for Additional Information Required for Clarification of Report)

LINE 17 – AUTHENTICATION ________________ (Report Authentication)

TITLE: FRIENDLY NUCLEAR STRIKE WARNING [STRIKWARN (NUC)]
REPORT NUMBER: F055 {USMTF # C505}

GENERAL INSTRUCTIONS: Use to provide friendly forces with the necessary information to take safety precautions against the effects of friendly nuclear blasts. Reference: ATP-45 (A), FM 3-3-1, and FM 3-4.

LINE 1 – DATE AND TIME	________________	(DTG)
LINE 2 – UNIT	________________	(Unit Making Report)
LINE 3 – ALFA WHISKEY	________________	(Target Number or Target Nickname)
LINE 4 – DELTA WHISKEY	________________	(DTG of Strike and DTG Strike Cancelled)
LINE 5 – FOXTROT ONE	________________	(Minimum Safe Distance 1 in Hundreds of Meters, Followed by Coordinates of Ground Zero or DGZ, or Coordinates Which Describe MSD Box ONE for Multiple Nuclear Detonations)
LINE 6 – FOXTROT TWO	________________	(Minimum Safe Distance 2 in Hundreds of Meters, Followed by Coordinates of Ground Zero or DGZ, or Coordinates Which Describe MSD Box TWO for Multiple Nuclear Detonations)
LINE 7 – FOXTROT THREE	________________	(Minimum Safe Distance 3 in Hundreds of Meters, Followed by Coordinates of Ground Zero or DGZ, or Coordinates Which Describe MSD Box THREE for Multiple Nuclear Detonations)
LINE 8 – GOLF	________________	(Means of Delivery/ROTA Classification)
LINE 9 – HOTEL WHISKEY	________________	(Number of Surface Bursts)
LINE 10 – INDIA WHISKEY	________________	(Number of Bursts if Multiple Strike)

**Continued on next page.

TITLE: FRIENDLY NUCLEAR STRIKE WARNING [STRIKWARN (NUC)] (CONTINUED)
REPORT NUMBER: F055 {USMTF # C505}

LINE 11 – NARRATIVE ______________________ (Free Text for Additional Information Required for Clarification of Report)

LINE 12 – AUTHENTICATION ______________________ (Report Authentication)

TITLE: GENERAL ADMINISTRATIVE MESSAGE [GENADMIN]
REPORT NUMBER: G001 {USMTF # F002}

GENERAL INSTRUCTIONS: Use to pass information not found in any other FM 101-5-2 report and message format. Use for free text message as required. Reference: FM 101-5.

LINE 1 – DATE AND TIME ______________________ (DTG)

LINE 2 – UNIT ______________________ (Unit Making Report)

LINE 3 – REFERENCE ______________________ (If Applicable, DTG of Referenced Report or Document)

LINE 4 – SUBJECT ______________________ (Subject Matter of This Report)

LINE 5 – NARRATIVE ______________________ (Free Text for Additional Information Required for Clarification of Report)

LINE 6 – AUTHENTICATION ______________________ (Report Authentication)

TITLE: HANDOVER MESSAGE |HANDOVER|
REPORT NUMBER: H001 {USMTF # E706}

GENERAL INSTRUCTIONS: Use to initiate and complete a handover from ADA command and control link to another. Reference: FM 44-100.

LINE 1 – DATE AND TIME ______________________ (DTG)

LINE 2 – UNIT ______________________ (Unit Making Report)

LINE 3 – CALL SIGN ______________________ (Call Sign of Aircraft or Flight to Handover)

LINE 4 – PRESENT POSITION ______________________ (Aircraft Position in UTM or Six-Digit Grid Coordinate With MGRS Grid Zone Designator)

LINE 5 – HEADING ______________________ (Heading in Degrees Magnetic)

LINE 6 – ALTITUDE ______________________ (Aircraft Altitude in Nearest 1,000 Feet)

LINE 7 – MISSION ______________________ (Mission Number of Aircraft/Flight)

LINE 8 – NO. TYPE ______________________ (Number and Type of Aircraft)

LINE 9 – SQUAWKING ______________________ (IFF/SIF Mode and Code)

LINE 10 – SPEED ______________________ (Speed in Knots if on an Intercept)

LINE 11 – ENGAGED ______________________ (Paired/Engaged Status if on an Intercept)

LINE 12 – WEAPON ______________________ (Weapon Status if on an Intercept)

LINE 13 – FUEL ______________________ (Fuel Remaining in Hundreds of Pounds)

LINE 14 – TIME TO BINGO ______________________ (Time in Hours and Minutes Until BINGO Fuel)

LINE 15 – TADIL ______________________ (TADIL Address)

**Continued on next page.

TITLE: HANDOVER REPORT [HANDOVER] (CONTINUED)
REPORT NUMBER: H001 {USMTF # E706}

LINE 16 – TARGET ____________	(Destination Target to Include Intermediate Points in UTM or Six-Digit Grid Coordinate With MGRS Grid Zone Designator)

**Lines 7 through 16 are amplifying information to be provided as desired.

LINE 17 – CONFIRM FREQUENCY ____________	(Frequency Designator)
LINE 18 – SEND ____________	(Call Sign of Aircraft or Flight Being Handed Over)
LINE 19 – FREQUENCY ____________	(Frequency Designator for Pilot to Contact Receiving Control Agency)
LINE 20 – ASSIGNED SQUAWK ____________	(IFF/SIF Mode and Code to be Assigned after Transfer)
LINE 21 – TIME ____________	(DTG that Handover is Effective)
LINE 22 – POSITION ____________	(UTM or Six-Digit Grid Coordinate With MGRS Grid Zone Designator)

**After the controlling agency contacts the pilot, he makes successful contact with the receiving agency; the receiving agency confirms transfer of control.

**The receiving agency passes the information in line 25 to the controlling agency.

____________THIS IS____________ HAVE CONTROL OF

LINE 23 – CALL SIGN ____________	(Call Sign of Aircraft or Flight Handed Over)
LINE 24 – NARRATIVE ____________	(Free Text for Additional Information Required for Clarification of Report)
LINE 25 – AUTHENTICATION ____________	(Report Authentication)

TITLE: HIGHWAY SITUATION REPORT [HWYSITREP]
REPORT NUMBER: H005 {USMTF # C200}

GENERAL INSTRUCTIONS: Use to provide data about main supply routes (MSRs) to include MSR capabilities, choke points, and units affected. Reference: FM 17-95 and FM 701-58.

LINE 1 – DATE AND TIME ________________ (DTG)

LINE 2 – UNIT ________________ (Unit Making Report)

LINE 3 – ROUTE ________________ (Main Supply Route Name and Number)

LINE 4 – EFFECTIVE ________________ (DTG Zone the Data is Effective)

LINE 5 – FROM ________________ (UTM or Six-Digit Grid Coordinate With MGRS Grid Zone Designator)

LINE 6 – TO ________________ (UTM or Six-Digit Grid Coordinate With MGRS Grid Zone Designator)

LINE 7 – DISTANCE ________________ (Distance in Kilometers Between the From and To Locations)

LINE 8 – NARRATIVE ________________ (Free Text for Additional Information Required for Clarification of Report)

LINE 9 – AUTHENTICATION ________________ (Report Authentication)

TITLE: HUMAN REMAINS SEARCH AND RECOVERY STATUS REPORT [REMAINSARSTAT]
REPORT NUMBER: H010 {USMTF # B965}

GENERAL INSTRUCTIONS: Use to report the status of searching for and recovering human remains within an assigned area of responsibility. Reference: FM 10-63-1 and FM 10-286.

LINE 1 – DATE AND TIME	______________	(DTG)
LINE 2 – UNIT	______________	(Unit Making Report)
LINE 3 – EFFECTIVE DTG	______________	(DTG the Information is Effective)
LINE 4 – STATUS	______________	(Enter BEGAN, COMPLETE, INCOMPLETE, DELAYED, HALTED, or PERFORMING to Indicate Search and Recovery Status)
LINE 5 – COORDINATES	______________	(UTM or Six-Digit Grid Coordinate With MGRS Grid Zone Designator)

**Send additional iterations of Line 5 as required. Number additional coordinate lines as 5a, 5b, 5c, and so on.

LINE 6 – RESULTS	______________	(Accomplishments Achieved From Search)
LINE 7 – REMAINS	______________	(Enter the Identification of the Remains Recovered)
LINE 8 – NARRATIVE	______________	(Free Text for Additional Information Required for Clarification of Report)
LINE 9 – AUTHENTICATION	______________	(Report Authentication)

TITLE: INTELLIGENCE REPORT [INTREP]
REPORT NUMBER: I001 {USMTF # C110}

GENERAL INSTRUCTIONS: Use to provide the exchange of information obtained through tactical collection efforts regarding events that could have an immediate and significant effect on current planning and operations of timely interest, such as the primary means of reporting HUMINT or counterintelligence information. Reference: FM 34-3.

LINE 1 – DATE AND TIME ______________________________ (DTG)

LINE 2 – UNIT ______________________________ (Unit Making Report)

LINE 3 – SIZE ______________________________ (Enemy Strength/Size/Number)

LINE 4 – ACTIVITY ______________________________ (Enemy Activity Description, Including Direction and Speed if Moving)

LINE 5 – LOCATION ______________________________ (UTM or Six-Digit Grid Coordinate With MGRS Grid Zone Designator)

LINE 6 – UNIT ______________________________ (Enemy Nationality, Unit Designator/Name/Type)

LINE 7 – TIME ______________________________ (DTG of Activity)

LINE 8 – EQUIPMENT ______________________________ (Major Enemy Equipment)

LINE 9 – SOURCES ______________________________ (Reliability Rating of Source and Credibility Rating of Information)

LINE 10 – EVALUATION ______________________________ (Evaluation of Source, Information, and BDA)

LINE 11 – CONCLUSION ______________________________ (Reporter's Analysis of What Reported Information Means)

LINE 12 – NARRATIVE ______________________________ (Free Text for Additional Information Required for Clarification of Report)

LINE 13 – AUTHENTICATION ______________________________ (Report Authentication)

TITLE: INTELLIGENCE SUMMARY [INTSUM]
REPORT NUMBER: I002

GENERAL INSTRUCTIONS: Use (1) to summarize significant enemy activities, (2) to report analysis of the current situation, and (3) to assess probable enemy courses of action in an area of operation. Send on O&I net as ROUTINE IAW unit SOPs or via liaison officer, courier, or digital routine precedence. Also, when sent graphically, this format represents a Graphic Summary. Use the Graphic Summary to graphically display (INSUM). Reference: FM 34-3.

LINE 1 – DATE AND TIME ______________________ (DTG)

LINE 2 – UNIT ______________________ (Unit Making Report)

LINE 3 – SITUATION ______________________ (General Enemy Situation Since Last Report (Deep, Close, Rear, Adjacent Units))

LINE 4 – EFLOT ______________________ (Current Enemy Front Line Trace)

LINE 5 – ENEMY UNIT SIZE______________________ (Enemy Ground Maneuver Units LOCATION/ACTIVITY/ Status by Echelon/Size, Location EST STRENGTH (Grid), Activity, Estimated Strength)

LINE 6 – ENEMY ARTILLERY ______________________ (Enemy Artillery Activity and Estimated Strength)

LINE 7 – ENEMY NBC ______________________ (Enemy NBC Activity (Type, Location, DTG))

LINE 8 – ENEMY AIR______________________ (Enemy Air and Air Activity)

LINE 9 – ENEMY ENGINEER ______________________ (Enemy Engineer Activity)

LINE 10 – REAR AREA THREAT ______________________ (Enemy Rear Area Threat (Light Forces, SF))

LINE 11 – ENEMY'S EST COA ______________________ (Enemy's Most Probable Courses of Action)

LINE 12 – PIR______________________ (Current PIR in Order of Priority and the Phase of Operation)

**Continued on next page.

TITLE: INTELLIGENCE SUMMARY [INSUM] (CONTINUED)
REPORT NUMBER: I002

LINE 13 – ENEMY CS ________	(Location and Activity of Enemy Combat Support Units)
LINE 14 – ENEMY CSS ________	(Location and Activity of Enemy Combat Service Support Units)
LINE 15 – VULNERABILITIES ________	(Analysis of Enemy's Current or Emerging Vulnerabilities)
LINE 16 – WEATHER AND TERRAIN ________	(Analysis of Effects of Weather and Terrain)
LINE 17 – ENEMY BDA ________	(Summarize Enemy BDA During Period)
LINE 18 – NARRATIVE ________	(Free Text for Additional Information Required for Clarification of Report)
LINE 19 – AUTHENTICATION ________	(Report Authentication)

TITLE: LOGISTICS RESUPPLY REQUEST [LOGRESREP]
REPORT NUMBER: L001

GENERAL INSTRUCTIONS: Use to convey to S4/G4 and logistical commander request of supplies and equipment prior to the next scheduled resupply. Use to request support in a tactical emergency or for urgent tactical requirements. Send normal or routine requirements through the LOGISTICS SITUATION REPORT. Use this format for all types of emergency or urgent supply requests (include water, barrier materiel, blade support, etc). Reference: FM 700-80 and FM 701-58.

LINE 1 – DATE AND TIME ______________________	(DTG)
LINE 2 – UNIT ______________________	(Unit Making Report)
LINE 3 – CLASS I ______________________	(Number of Meals or Water Required Before Next Scheduled Resupply. State Precedence.)
LINE 4 – CLASS III ______________________	(Type and Amount of POL Required Before Next Resupply. State Precedence.)
LINE 5 – CLASS V ______________________	(Type and Amount of Ammunition Required Before Next Resupply. State Precedence.)
LINE 6 – MAINTENANCE ______________________	(Number and Quantity by Line Number of Parts and/or Equipment Required Before Next Resupply. State Precedence.)
LINE 7 – MEDICAL ______________________	(Number and Type of Medical Equipment/Medical Support Required Before Next Resupply. State Precedence.)
LINE 8 – OTHER ______________________	(Number and Type of Supplies Not Listed Above and Required Before the Next Resupply. State Precedence.)
LINE 9 – RESUPPLY LOCATION ______________________	(UTM or Six-Digit Grid Coordinate With Grid Zone Designator for Location of Resupply Requested)

**Continued on next page.

TITLE: LOGISTICS RESUPPLY REQUEST [LOGRESREP] (CONTINUED)
REPORT NUMBER: L001

LINE 10 – SUPPLY STAT ______________________	(Status of UNIT (Red, Amber, Green))
LINE 11 – NARRATIVE ______________________	(Free Text for Additional Information Required for Clarification of Report)
LINE 12 – AUTHENTICATION ______________________	(Report Authentication)

TITLE: LOGISTICS SITUATION REPORT [LOGSITREP]
REPORT NUMBER: L005 {USMTF # C802}

GENERAL INSTRUCTIONS: Use to report logistic problems, required logistic assistance and reallocation, and recommended or intended courses of actions. Reference: FM 700-80 and FM 701-58.

LINE 1 – DATE AND TIME ______________________________ (DTG)

LINE 2 – UNIT ______________________________ (Unit Making Report)

LINE 3 – MAP ______________________________ (UTM or Six-Digit Grid Coordinate With MGRS Grid Zone Designator for the Map Being Referenced)

LINE 4 – EVALUATION ______________________________ (Major Units, Combat Operation Type, and Days Logistically Supportable)

UNIT________________OPERATION_________________DAYS_________

UNIT________________OPERATION_________________DAYS_________

UNIT________________OPERATION_________________DAYS_________

LINE 5 – IDENTIFICATION ______________________________ (Identification/Designation of Component, Agency, or Unit Making Report)

LINE 6 – LOCATION ______________________________ (UTM or Six-Digit Grid Coordinate With MGRS Grid Zone Designator)

LINE 7 – COMMENT ______________________________ (Pertinent Unit Comments)

LINE 8 – FORCE WEAPON SYSTEMS ______________________ (Line Item Numbers and Quantities on Hand, Authorized, and Operational)

LINE___________ON HAND________AUTH_______OPER_____REQ_____

LINE___________ON HAND________AUTH_______OPER_____REQ_____

LINE___________ON HAND________AUTH_______OPER_____REQ_____

**Continued on next page.

TITLE: LOGISTICS SITUATION REPORT [LOGSITREP] (CONTINUED)
REPORT NUMBER: L005 {USMTF # C802}

LINE 9 – AMMUNITION ________________________________ (Type of Ammunition, Quantity on Hand, Quantity Due in, and RDD)

TYPE____________ON HAND_________DUE____________RDD_____ REQ_____

TYPE____________ON HAND_________DUE____________RDD_____ REQ_____

TYPE____________ON HAND_________DUE____________RDD_____ REQ_____

LINE 10 – CRYOGENICS ________________________________ (Comments or Status of Cryogenics)

LINE 11 – PETROLEUM ________________________________ (Product Type, Quantity and Unit of Measure on Hand, and Days of Supply on Hand)

PRODUCT____________ON HAND____________DAYS___________REQ_____

PRODUCT____________ON HAND____________DAYS___________REQ_____

LINE 12 – EQUIPMENT ________________________________ (National Stock Numbers and Quantities on Hand, Authorized, and Operational)

NSN_______________ON HAND______AUTH________OPER________REQ_____

NSN_______________ON HAND______AUTH________OPER________REQ_____

NSN_______________ON HAND______AUTH________OPER________REQ_____

LINE 13 – CRITICAL EQUIPMENT ________________________ (Type, Quantity Nonoperational, Materiel Damage Level (MDLV) (MODERATE or SEVERE), and as of DTG)

TYPE_______________NONOP______DMLV________AS OF________REQ_____

TYPE_______________NONOP______DMLV________AS OF________REQ_____

TYPE_______________NONOP______DMLV________AS OF________REQ_____

**Continued on next page.

TITLE: LOGISTICS SITUATION REPORT [LOGSITREP] (CONTINUED)
REPORT NUMBER: L005 {USMTF # C802}

LINE 14 – REPAIR MATERIEL ______________________ (Type, National Stock Number, and Quantity of Critically Required Repair Materiel)

TYPE__________ NSN__________________ QTY__________ REQ____

TYPE__________ NSN__________________ QTY__________ REQ____

TYPE__________ NSN__________________ QTY__________ REQ____

LINE 15 – RATIONS ______________________ (Ration Type, Number of Meals on Hand, and Days of Supply on Hand)

TYPE__________ ON HAND________________ DAYS________ REQ____

TYPE__________ ON HAND________________ DAYS________ REQ____

TYPE__________ ON HAND________________ DAYS________ REQ____

LINE 16 – WATER______________________ (Water Type (Potable or Nonpotable) Gallons Available, and Days of Supply on Hand)

TYPE__________ GALLONS________________ DAYS________ REQ____

TYPE__________ GALLONS________________ DAYS________ REQ____

TYPE__________ GALLONS________________ DAYS________ REQ____

**Repeat lines 5 through 16 to indicate each type of equipment. Assign sequential line numbers to succeeding iterations; for example, first iteration 5 through 16; second iteration 5a through 16a; third iteration 5b through 16b; and so on.

LINE 17 – CDR'S SUMMARY ______________________ (Narrative Comments Addressing Degraded Capabilities, Shortages, or Problems Impacting on Warfighting Capability as well as Recommended Courses of Action)

LINE 18 – NARRATIVE ______________________ (Free Text for Additional Information Required for Clarification of Report)

LINE 19 – AUTHENTICATION ______________________ (Report Authentication)

TITLE: LOST SENSITIVE ITEM REPORT [LOSTITEM]
REPORT NUMBER: L010

GENERAL INSTRUCTIONS: Use—usually by the Provost Marshall—to summarize unit's loss of sensitive items. Send IAW unit SOPs. References FM 19-1.

LINE 1 – DATE AND TIME ______________________________ (DTG)

LINE 2 – UNIT ______________________________ (Unit Making Report)

LINE 3 – SUMMARY ______________________________ (Summary of Loss)

UNIT	ITEMS LOST	SERIAL NO.	ACTION TAKEN	DTG LOSS
________	________	________	________	________
________	________	________	________	________
________	________	________	________	________
________	________	________	________	________

LINE 4 – PM ASSESSMENT ______________________________

LINE 5 – NARRATIVE ______________________________ (Free Text for Additional Information Required for Clarification of Report)

LINE 6 – AUTHENTICATION ______________________________ (Report Authentication)

TITLE: MAIL DISTRIBUTION SCHEME CHANGE [MAILDISTCH]
REPORT NUMBER: M001 {USMTF # F882}

GENERAL INSTRUCTIONS: Use to establish and change mail routing for assigned or co-located units. Reference: FM 12-6.

LINE 1 – DATE AND TIME	______________	(DTG)
LINE 2 – UNIT	______________	(Unit Making Report)
LINE 3 – GAINING NUMBER	______________	(Number of the Facility Gaining Responsibility)
LINE 4 – GAINING LOCATION	______________	(Place Name for the Facility Gaining Responsibility)
LINE 5 – EFFECTIVE DATE	______________	(DTG the Change in Mail Distribution Will be Effective)
LINE 6 – LOSING NUMBER	______________	(Number of the Facility Losing Responsibility)
LINE 7 – LOSING LOCATION	______________	(Place Name for the Facility Losing Responsibility)
LINE 8 – EFFECTIVE DATE	______________	(DTG the Change in Losing Facility Will be Effective)
LINE 9 – FOOTNOTE CODES	______________	(Letter(s) or Letter-Number(s) of Applicable Postal Footnote(s))
LINE 10 – GAINING FACILITY	______________	(Location Name of the Affected Military Post Office)
LINE 11 – MAIL CLASS	______________	(Mail Classification Code(s) Permitted at the Location Named in Line 9 (if not ALL, Also Provide Required Disposition of Mail Classifications not Permitted))
LINE 12 – NARRATIVE	______________	(Free Text for Additional Information Required for Clarification of Report)
LINE 13 – AUTHENTICATION	______________	(Report Authentication)

TITLE: MAINTENANCE SUPPORT REQUEST [MAINTSPTREQ]
REPORT NUMBER: M005 {USMTF # D840}

GENERAL INSTRUCTIONS: Use to request maintenance support. Reference: FM 9-43-1.

LINE 1 – DATE AND TIME ______________________________ (DTG)

LINE 2 – UNIT ______________________________ (Unit Making Report)

LINE 3 – REQUESTING AGENCY ______________________________ (Identifier or Designator of the Requesting Agency)

LINE 4 – LOCATION ______________________________ (UTM or Six-Digit Grid Coordinate With MGRS Grid Zone Designator)

LINE 5 – DATE ______________________________ (Date Maintenance Support is Required)

LINE 6 – NOMENCLATURE ______________________________ (Category, Name, and Model of Equipment Requiring Maintenance Support)

LINE 7 – NO. PIECES ______________________________ (Number of Pieces of Equipment Requiring Maintenance Support)

LINE 8 – TYPE______________________________ (Type of Maintenance Support Required)

**Repeat lines 3 through 8 for each type of equipment requiring maintenance support. Assign sequential line numbers to succeeding iterations; for example, first iteration 3 through 8; second iteration 3a through 8a, and so on.

LINE 9 – EQUIPMENT LOCATION ______________________________ (Location of Equipment Needing Support Using UTM or Six-Digit Grid Coordinate With Grid Zone Designator)

LINE 10 – CONDITION______________________________ (Brief Description of Equipment Condition)

LINE 11 – SPECIAL INSTRUCTIONS______________________________ (Special Instructions Relevant to Special Tools, Components Needed to Perform Support)

**Continued on next page.

TITLE: MAINTENANCE SUPPORT REQUEST [MAINTSPTREQ] (CONTINUED)
REPORT NUMBER: M005 {USMTF # D840}

LINE 12 – COORDINATION ______________	(Coordination Instructions Required Between Supported and Supporting Units)
LINE 13 – NARRATIVE ______________	(Free Text for Additional Information Required for Clarification of Report)
LINE 14 – AUTHENTICATION ______________	(Report Authentication)

TITLE: MAINTENANCE SUPPORT RESPONSE [MAINTSPTRES]
REPORT NUMBER: M010 {USMTF # D841}

GENERAL INSTRUCTIONS: Use to reply to a request for maintenance support. Reference: FM 9-43-1.

LINE 1 – DATE AND TIME ________________ (DTG)

LINE 2 – UNIT ________________ (Unit Making Report)

LINE 3 – REFERENCE ________________ (DTG and Other Identifying Information of Communications Which Requested Maintenance Support)

LINE 4 – DISPOSITION ________________ (Status of Request: APPROVED, DISAPPROVED, MODIFIED)

LINE 5 – SUPPORTING UNIT ________________ (Identifier or Designator of Supporting Unit)

LINE 6 – LOCATION ________________ (UTM or Six-Digit Grid Coordinate With MGRS Grid Zone Designator)

LINE 7 – PRIORITY ________________ (Priority of Maintenance Support Mission)

LINE 8 – START ________________ (DTG the Requested Support Will Start)

LINE 9 – COMPLETE ________________ (Estimated DTG Support Mission Will be Completed)

LINE 10 – INSTRUCTIONS ________________ (Special Instructions Relevant to Support Required by the Supporting Unit)

LINE 11 – COORDINATION ________________ (Coordination Instructions Required Between Supporting Units)

**Lines 5 through 11 are applicable only if line 4 is approved or modified.

LINE 12 – NARRATIVE ________________ (Free Text for Additional Information Required for Clarification of Report)

LINE 13 – AUTHENTICATION ________________ (Report Authentication)

TITLE: MAJOR AMMUNITION MALFUNCTION-INITIAL REPORT [AMMOMALFUNCREP]
REPORT NUMBER: M015 {USMTF # C201}

GENERAL INSTRUCTIONS: Use to provide initial notification (report) of a major ammunition malfunction. Reference: FM 9-6.

LINE 1 – DATE AND TIME ______________________ (DTG)

LINE 2 – UNIT ______________________ (Unit Making Report)

LINE 3 – ORIGINATING UNIT ______________________ (Identifier for the Unit Making the Report)

LINE 4 – DATE-TIME ______________________ (DTG at Which the Incident Occurred)

LINE 5 – LOCATION STATUS ______________________ (Status of the Location at Which the Incident Occurred)

LINE 6 – INSTALLATION TYPE ______________________ (Type of Installation or Location at Which Incident Occurred)

LINE 7 – LOCATION DESIGNATOR ______________________ (UTM or Six-Digit Grid Coordinate With MGRS Grid Zone Designator of the Incident)

LINE 8 – COUNTRY ______________________ (Country or Geographical-Political Pact of the Organization Involved in the Incident)

LINE 9 – UTM ______________________ (UTM or Six-Digit Grid Coordinate With MGRS Grid Zone Designator)

**Repeat line 9 as often as necessary to enter the UTM grid reference points.

LINE 10 – MALFUNCTION ______________________ (Enter Details and Cause of Malfunction (See NOTE 1))

LINE 11 – DAMAGES/CASUALTIES ______________________ (Casualty and Damage Information)

**Continued on next page.

TITLE: MAJOR AMMUNITION MALFUNCTION-INITIAL REPORT [AMMOMALFUNCREP] (CONTINUED)
REPORT NUMBER: M015 {USMTF # C201}

LINE 12 – STOCK NUMBER/CODE ____________________ (National Stock Number, NATO Stock Number or National Short Code)

LINE 13 – BATCH ____________________ (Batch Number of the Ammunition)

LINE 14 – LOT ____________________ (Lot Number of the Ammunition)

LINE 15 – WEAPON ____________________ (Identifier of the Generic Type of Weapon Being Used)

LINE 16 – MODEL ____________________ (Exact Weapon Model)

LINE 17 – NUMBER ____________________ (Number of Weapons Involved)

**Repeat lines 15 through 17 for each type of weapon. Assign sequential line numbers to succeeding iterations; for example, first iteration 15 through 17; second iteration 15a through 17a; third iteration 15b through 17b; and so on.

LINE 18 – WEATHER ____________________ (Weather Conditions at Time of Malfunction)

LINE 19 – RANK ____________________ (Rank or Position of POC)

LINE 20 – SURNAME ____________________ (Surname of POC)

LINE 21 – UNIT ____________________ (Unit POC)

LINE 22 – TELEPHONE ____________________ (Telephone Number of POC)

LINE 23 – ACTION TAKEN ____________________ (Enter Information on Action Taken)

LINE 24 – NARRATIVE ____________________ (Free Text for Additional Information Required for Clarification of Report)

LINE 25 – AUTHENTICATION ____________________ (Report Authentication)

TITLE: MEACONING, INTRUSION, JAMMING, AND INTERFACE (MIJI) FEEDER REPORT [MIJIFEEDER]
REPORT NUMBER: M020 {USMTF # C120}

GENERAL INSTRUCTIONS: Use to share MIJI incidents in a timely manner and to provide for joint exchange of tactical MIJI information including electro-optic interference. Reference: FM 34-10.

LINE 1 – DATE AND TIME ______________________ (DTG)

LINE 2 – UNIT ______________________ (Unit Making Report)

LINE 3 – INTERFERENCE ______________________ (Strength and Characteristics)

LINE 4 – LOCATION ______________________ (UTM of Affected Station or Six-Digit Grid Coordinate With MGRS Grid Zone Designator)

LINE 5 – ON TIME ______________________ (Start DTG)

LINE 6 – OFF TIME ______________________ (End DTG)

LINE 7 – EFFECTS ______________________ (Operations/Equipment Affected)

LINE 8 – FREQUENCY ______________________ (Frequency/Frequency Range Affected)

LINE 9 – NARRATIVE ______________________ (Free Text for Additional Information Required for Clarification of Report)

LINE 10 – AUTHENTICATION ______________________ (Report Authentication)

TITLE: MEDIA CONTACT REPORT [MEDIACOTREP]
REPORT NUMBER: M025

GENERAL INSTRUCTIONS: Use to inform commander and staff of noncredentialed, unescorted, or unregistered media in the Division AO. Reference: FM 46-1.

LINE 1 – DATE AND TIME ______________________________ (DTG)

LINE 2 – UNIT ______________________________ (Unit Making Report)

LINE 3 – NONACCREDITED MEDIA______________________ (Number of Any Unescorted, Noncredentialed or Unregistered Media Representatives in AO)

LINE 4 – UNESCORTED MEDIA REPRESENTITIVE____________

a. N/UMR ______________________________ (Number Unescorted)

b. MEDIA AFFILIATION______________________ (Country/Agency)

c. LOCATION______________________________ (UTM or Six-Digit Grid Coordinate With MGRS Grid Zone Designator)

d. REASON FOR DETAINING ____________________ (Unescorted, Noncredentialed, Unregistered)

e. SOURCE ______________________________ (Large Circulation Newspaper/ Periodical or National Television/YES or NO)

f. DURATION______________________________ (Duration in Location in Days (YES or NO))

**Repeat lines 4a through 4f to report additional unescorted representatives. Assign sequential line numbers to succeeding iterations; for example, first iteration 4a through 4f; second iteration 4a1 through 4f1; third iteration 4a2 through 4f2; and so on.

LINE 5 – PAO ASSESSMENT ______________________ (Public Affairs Officer's Report)

LINE 6 – NARRATIVE ___________________________ (Free Text for Additional Information Required for Clarification of Report)

LINE 7 – AUTHENTICATION ______________________ (Report Authentication)

TITLE: MEDICAL EVACUATION REQUEST [MEDEVAC]
REPORT NUMBER: M030 {USMTF # D901}

GENERAL INSTRUCTIONS: Use to request evacuation of sick and wounded personnel by other than USAF fixed-wing assets. Reference: FM 8-10-6.

LINE 1 – DATE AND TIME ______________________	(DTG)
LINE 2 – UNIT ______________________	(Unit Making Report)
LINE 3 – LOCATION ______________________	(UTM or Six-Digit Grid Coordinate With MGRS Grid Zone Designator of Pickup Site Location; Encrypt if Using Non-Secure Means of Communication)
LINE 4 – RADIO FREQUENCY ______________________	(Radio Frequency at Pickup Site; Call Sign, and Suffix)
LINE 5 – NUMBER OF PATIENTS(P)______________________	(Number of Patients by Precedence. Report Only Applicable Information and Encrypt the Brevity Codes: A – URGENT B – URGENT-SURG C – PRIORITY D – ROUTINE E – CONVENIENCE) If two or more categories must be reported in the same request, insert the word BREAK between each category.)
LINE 6 – SPECIAL EQUIPMENT______________________	(Encrypt the Applicable Brevity Codes. A – None B – Hoist C – Extraction Equipment)
LINE 7 – NUMBER OF PATIENTS(T)______________________	(Total Number of Patients by Type. L + # of PNT – Litter A + # of PNT – Ambulatory)

**Continued on next page.

TITLE: MEDICAL EVACUATION REQUEST [MEDEVAC] (CONTINUED)
REPORT NUMBER: M030 {USMTF # D901}

LINE 8 – SECURITY OF PICK-UP SITE ______________________ (N – No Enemy Troops in Area
P – Possibly Enemy Troops in Area (approach with caution)
E – Enemy Troops in Area (approach with caution)
X – Enemy Troops in Area (armed escort required))

LINE 9 – NO. TYPE OF WOUND, INJURY, OR ILLNESS (PEACETIME)

____ (Specific Information Regarding Patient Wounds by Type)

LINE 10 – METHOD OF MARKING AT PZ ___________________ (Encrypt the Brevity Codes:
A – Panels
B – Pyrotechnic signal
C – Smoke Signal
D – None
E – Other)

LINE 11 – PATIENT NATIONALITY ______________________ (Number of Patients in Each Category Need Not be Transmitted. Encrypt Only the Applicable Brevity Codes:
A – US Military
B – US Civilian
C – Non-US Military
D – Non-US Civilian
E – EPW)

LINE 12 – NBC CONTAMINATION _______________________ (NBC Contamination; Use Brevity Codes:
N = Nuclear
B = Biological
C = Chemical
U = Unknown
A = All Clear)

LINE 13 – TERRAIN DESCRIPTION_______________________ (Description of Terrain Features at the Proposed Pick-up Site)

LINE 14 – AUTHENTICATION ___________________________ (Report Authentication)

TITLE: MEDICAL SITREP [MEDSITREP]
REPORT NUMBER: M035

GENERAL INSTRUCTIONS: Use to convey a quick, consolidated medical status report as a snapshot rather than a full report. Reference: FM 8-10-3.

LINE 1 – DATE AND TIME	______________	(DTG)
LINE 2 – UNIT	______________	(Unit Making Report)
LINE 3 – LOCATION	______________	(UTM or Six-Digit Grid Coordinate With MGRS Grid Zone Designator)
LINE 4 – STATUS	______________	(Number of Operational Cots, Unoccupied Cots, Number of Cots Uploaded)
LINE 5 – PATIENTS	______________	(Number of Patients)
LINE 6 – PROJECTED LOCATIONS	______________	(Anticipated Operations in the Next 24 Hours; Anticipated Opening and Closing Times at New Locations)
LINE 7 – COMBAT HEALTH LOGISTICS	______________	(Number of Days Class VIII & Blood OH)
LINE 8 – EVACUATION ASSETS	______________	(Availability and Operational Capacity of Evacuation Assets)
LINE 9 – NARRATIVE	______________	(Free Text for Additional Information Required for Clarification of Report)
LINE 10 – AUTHENTICATION	______________	(Report Authentication)

TITLE: MEDICAL SPOT REPORT [MEDSPTREP]
REPORT NUMBER: M040

GENERAL INSTRUCTIONS: Use to report major medical incidents. Reference: FM 8-10-1.

LINE 1 – DATE AND TIME ______________________ (DTG)

LINE 2 – UNIT ______________________ (Unit Making Report)

LINE 3 – INCIDENT ______________________ (DTG of Incident)

LINE 4 – LOCATION ______________________ (UTM or Six-Digit Grid Coordinate With MGRS Grid Zone Designator of Emergency)

LINE 5 – EMERGENCY ______________________ (Specified Medical Emergency)

LINE 6 – UNIT AFFECTED ______________________ (Affected Friendly Unit)

LINE 7 – NARRATIVE ______________________ (Free Text for Additional Information Required for Clarification of Report)

LINE 8 – AUTHENTICATION ______________________ (Report Authentication)

TITLE: MEDICAL STATUS REPORT [MEDSTAT]
REPORT NUMBER: M045 {USMTF # B907}

GENERAL INSTRUCTIONS: Use to provide status on hospitalization, incidence or occurrence of disease, and unresolved problems or items of significant interest having impact upon the overall medical capability or health of the command. The MEDSTAT report must be repeated in its entirety to report the medical status of ashore or afloat units. Reference: FM 8-10-24.

LINE 1 – DATE AND TIME ________________________ (DTG)

LINE 2 – UNIT ________________________ (Unit Making Report)

LINE 3 – TYPE ________________________ (Type of Medical Status Report: ASHORE REPORT or AFLOAT REPORT)

LINE 4 – COMMAND ________________________ (Name of Designator of Reporting Command or Facility if Ashore; Ship Name, Type and Hull Number)

LINE 5 – POC ________________________ (Name of Medical Point of Contact)

LINE 6 – AVAILABLE ________________________ (Number of Cots or Hospital Beds Available)

LINE 7 – OCCUPIED ________________________ (Number of Cots or Hospital Beds Occupied)

LINE 8 – OVERFLOW ________________________ (Number of Cots/Holding Beds Occupied)

LINE 9 – FULL ________________________ (Number of Cots/Holding Beds Occupied)

LINE 10 – WAITING ________________________ (Number of Patients Awaiting Evacuation Out of/From Command's Area of Responsibility Since Last Report)

LINE 11 – MEDLOG ________________________ (Number of Days of Class VIII Supplies on Hand)

LINE 12 – PROBLEMS ________________________ (Unresolved Problems and Anticipated Problems/Issues Affecting Reporting Command)

**Continued on next page.

TITLE: MEDICAL STATUS REPORT [MEDSTAT] (CONTINUED)
REPORT NUMBER: M045 {USMTF # B907}

LINE 13 – ASSESSMENT ______________________ (Commanding Officer's Assessment of Operational Capabilities and Ability to Continue Medical Support (Mandatory Entry))

LINE 14 – NARRATIVE ______________________ (Free Text for Additional Information Required for Clarification of Report)

**Report lines 3 to 14 as a group when reporting the medical status of more than one command, facility, or ship.

LINE 15 – AUTHENTICATION ______________________ (Report Authentication)

TITLE: MESSAGE CORRECTION/CANCELLATION [MSGCORRN]
REPORT NUMBER: M050 {USMTF # C002}

GENERAL INSTRUCTIONS: Use to cancel a message, add, delete, or replace information in the referenced message by a message originator. Reference: FM 11-32.

LINE 1 – DATE AND TIME ________________ (DTG)

LINE 2 – UNIT ________________ (Unit Making Report)

LINE 3 – REPORT ________________ (Report Type Referenced)

LINE 4 – ORIGINATOR ________________ (Originator of Referenced Report or Order)

LINE 5 – SERIAL ________________ (Serial Number of Referenced Report or Order)

LINE 6 – ACTION ________________ (Action Directed: CANCEL, ADD, DELETE, or REPLACE)

LINE 7 – NARRATIVE ________________ (Free Text for Additional Information Required for Clarification of Report)

LINE 8 – AUTHENTICATION ________________ (Report Authentication)

TITLE: METEOROLOGICAL-COMPUTER MESSAGE [MET.CM]
REPORT NUMBER: M055 {USMTF # F254}

GENERAL INSTRUCTIONS: Use to disseminate computer meteorological data for use in fire support technical fire control computations. Reference: FM 6-20 and FM 6-16 through FM 6-16-3.

LINE 1 – DATE AND TIME ________________________________ (DTG)

LINE 2 – UNIT ____ (Unit Making Report)

(GLOBAL OCTANT) (MET STA POSITION OR NAME) (MET VALIDITY) (MET STA HEIGHT & PRESSURE)

LINE 3 – ALFA ____

(MET.CM ALTITUDE ZONE) (MET.CM WIND DIR AND SPEED) (MET.CM AIR VIRTUAL TEMP & PRESSURE)

Sub-Line 00____

Sub-Line 01____

Sub-Line 02____

Sub-Line 03____

Sub-Line 04____

Sub-Line 05____

(MET.CM ALTITUDE ZONE) (MET.CM WIND DIR AND SPEED) (MET.CM AIR VIRTUAL TEMP & PRESSURE)

Sub-Line 06____

Sub-Line 07____

Sub-Line 08____

Sub-Line 09____

Sub-Line 10____

Sub-Line 11____

Sub-Line 12____

Sub-Line 13____

** Continued on next page.

TITLE: METEOROLOGICAL-COMPUTER MESSAGE [MET.CM] (CONTINUED)
REPORT NUMBER: M055 {USMTF # F254}

Sub-Line 14____

Sub-Line 15____

LINE 4 – NARRATIVE ______________________ (Free Text for Additional Information Required for Clarification of Report)

LINE 5 – AUTHENTICATION ______________________ (Report Authentication)

TITLE: METEOROLOGICAL-FALLOUT MESSAGE [MET.CF]
REPORT NUMBER: M060 {USMTF # F251}

GENERAL INSTRUCTIONS: Use to disseminate fallout meteorological data used in computing fallout predictions and in chemical planning. Reference: FM 6-20 and FM 6-16 through FM 6-16-3

LINE 1 – DATE AND TIME ________________________________ (DTG)

LINE 2 – UNIT ____ (Unit Making Report)

(GLOBAL OCTANT) (MET STA POSITION OR NAME) (PERIOD OF VALIDITY) (MET STA HEIGHT & PRESSURE)

LINE 3 – ALFA ____

(MET.CF ALTITUDE ZONE) (MET WIND DIRECTION TENS OF MILS) (MET.CF WIND SPEED IN KNOTS)

Sub-Line 00____

Sub-Line 01____

Sub-Line 02____

Sub-Line 03____

Sub-Line 04____

(MET.CF ALTITUDE ZONE) (MET WIND DIRECTION TENS OF MILS) (MET.CF WIND SPEED IN KNOTS)

Sub-Line 05____

Sub-Line 06____

Sub-Line 07____

Sub-Line 08____

Sub-Line 09____

Sub-Line 10____

Sub-Line 11____

Sub-Line 12____

Sub-Line 13____

**Continued on next page.

TITLE: METEOROLOGICAL-FALLOUT MESSAGE [MET.CF] (CONTINUED)
REPORT NUMBER: M060 {USMTF # F251}

Sub-Line 14____

Sub-Line 15____

LINE 4 – NARRATIVE __________________________ (Free Text for Additional Information Required for Clarification of Report)

LINE 5 – AUTHENTICATION ______________________ (Report Authentication)

TITLE: METEOROLOGICAL-TARGET ACQUISITION MESSAGE [MET.TA]
REPORT NUMBER: M065 {USMTF # F255}

GENERAL INSTRUCTIONS: Use to disseminate meteorological data for use by target acquisition systems. Reference: FM 6-20 and FM 6-16 through FM 6-16-3.

LINE 1 – DATE AND TIME ______________________________ (DTG)

LINE 2 – UNIT ____ (Unit Making Report)

(GLOBAL OCTANT) (MET STA POSITION OR NAME) (MET VALIDITY) (MET STA HEIGHT & PRESSURE)

LINE 3 – ALFA ____

(CLOUD BASE HEIGHT) (MEAN REFRACTIVE INDEX)

LINE 4 – BRAVO ____

(MET.TA ALTITUDE ZONE) (MET.TA WIND DIR & SPEED) (MET.TA AIR AMBIENT TEMP & RELATIVE HUMIDITY)

Sub-Line 00____

Sub-Line 01____

(MET.TA ALTITUDE ZONE) (MET.TA WIND DIR & SPEED) (MET.TA AIR AMBIENT TEMP & RELATIVE HUMIDITY)

Sub-Line 02____

Sub-Line 03____

Sub-Line 04____

Sub-Line 05____

Sub-Line 06____

Sub-Line 07____

Sub-Line 08____

Sub-Line 09____

Sub-Line 10____

**Continued on next page.

TITLE: METEOROLOGICAL-TARGET ACQUISITION MESSAGE [MET.TA] (CONTINUED)
REPORT NUMBER: M065 {USMTF # F255}

Sub-Line 11____

Sub-Line 12____

Sub-Line 13____

Sub-Line 14____

Sub-Line 15____

LINE 5 – NARRATIVE ________________________ (Free Text for Additional Information Required for Clarification of Report)

LINE 6 – AUTHENTICATION ____________________ (Report Authentication)

TITLE: MILITARY POSTAL FACILITY REQUEST [POSTREQ]
REPORT NUMBER: M070 {USMTF # D881}

GENERAL INSTRUCTIONS: Use to request authorization to establish or disestablish a military postal facility. Reference: FM 16-1.

LINE 1 – DATE AND TIME ______________________	(DTG)
LINE 2 – UNIT ______________________	(Unit Making Report)
LINE 3 – TYPE FACILITY ______________________	(Identify the Type Postal Facility Reported)
LINE 4 – DATE (OPENING) (CLOSING) ______________________	(State Whether Proposed Opening or Proposed Closing and the DTG)
LINE 5 – PERSONNEL ______________________	(Number of Personnel Affected)
LINE 6 – UNITS ______________________	(Major Units Affected)
LINE 7 – LOCATION ______________________	(Enter the Place Name of the Current or Proposed Postal Facility, the UTM or Six-Digit Grid Coordinate With MGRS Grid Zone Designator)
LINE 8 – JUSTIFICATION ______________________	(Information to Rationalize the Need to Open or Close a Military Postal Facility)
LINE 9 – NARRATIVE ______________________	(Free Text for Additional Information Required for Clarification of Report)
LINE 10 – AUTHENTICATION ______________________	(Report Authentication)

TITLE: MODIFICATION-ATTACK CRITERIA [MOD.ATTACK]
REPORT NUMBER: M075 {USMTF # E277}

GENERAL INSTRUCTIONS: Use to override desired effects or standard volley factors in the Joint Munitions Effectiveness Manual [JMEM] for specified target types and subtypes. Reference: FM 6-20.

LINE 1 – DATE AND TIME ______________________________ (DTG)

LINE 2 – UNIT ______________________________ (Unit Making Report)

LINE 3 – PLAN ______________________________ (Fire Plan Designation)

LINE 4 – POI ______________________________ (Primary Option Indicator. Enter ADD, AMEND, or CANCEL)

LINE 5 – TARGET ______________________________ (Target Type and Subtype)

LINE 6 – PROTECTION ______________________________ (Degree of Personnel Protection)

LINE 7 – EFFECTS ______________________________ (Effects (Percent Damage) Required)

LINE 8 – VOLLEYS ______________________________ (Fire for Effect Number of Volleys)

LINE 9 – NARRATIVE ______________________________ (Free Text for Additional Information Required for Clarification of Report)

LINE 10 – AUTHENTICATION ______________________________ (Report Authentication)

TITLE: MODIFICATION-EXCLUDE CRITERIA [MOD.XCLUDE]
REPORT NUMBER: M080 {USMTF # E276}

GENERAL INSTRUCTIONS: Use to disseminate exclusions from tactical fire control consideration for fire missions or specified fire-plan processing. Exclusions include specific weapons of fire units or combinations of shell and fuze by weapon type or fire unit. Reference: FM 6-20 and FM 6-16 through FM 6-16-3.

LINE 1 – DATE AND TIME ________________ (DTG)

LINE 2 – UNIT ________________ (Unit Making Report)

LINE 3 – PLAN ________________ (Fire Plan Designation)

LINE 4 – TIME ________________ (DTG of Fire Plan)

LINE 5 – POI ________________ (Primary Option Indicator. Enter ADD, AMEND, or CANCEL)

LINE 6 – EXCLUDE ________________ (Section, Platoon, Battery/Company, Battalion/Regiment, and Regiment/Brigade/Division Designators, or Ship Call Sign(s))

LINE 7 – EXCLUDE ________________ (Section, Platoon, Battery/Company, Battalion/Regiment, and Regiment/Brigade/Division Designators, or Ship Call Sign(s))

LINE 8 – EXCLUDE ________________ (Section, Platoon, Battery/Company, Battalion/Regiment, and Regiment/Brigade/Division Designators, or Ship Call Sign(s))

LINE 9 – EXCLUDE ________________ (Section, Platoon, Battery/Company, Battalion/Regiment, and Regiment/Brigade/Division Designators, or Ship Call Sign(s))

LINE 10 – EXCLUDE ________________ (Section, Platoon, Battery/Company, Battalion/Regiment, and Regiment/Brigade/Division Designators, or Ship Call Sign(s))

**Continued on next page

TITLE: MODIFICATION-EXCLUDE CRITERIA [MOD.XCLUDE] (CONTINUED)
REPORT NUMBER: M080 {USMTF # E276}

LINE 11 – WEAPON EXCLUDED ______________________ (Surface-to-Surface Weapon Type)

LINE 12 – WEAPON ______________________ (Surface-to-Surface Weapon Type)

LINE 13 – PROJECTILE EXCLUDED ______________________ (Projectile Excluded)

LINE 14 – PROJECTILE ______________________ (Projectile Excluded)

LINE 15 – PROPELLANT EXCLUDED ______________________ (Propellant Excluded)

LINE 16 – PROPELLANT ______________________ (Propellant Excluded)

LINE 17 – FUZE EXCLUDED ______________________ (Fuze Excluded)

LINE 18 – FUZE ______________________ (Fuze Excluded)

LINE 19 – NARRATIVE ______________________ (Free Text for Additional Information Required for Clarification of Report)

LINE 20 – AUTHENTICATION ______________________ (Report Authentication)

TITLE: NBC SITUATION REPORT [NBCSITREP]
REPORT NUMBER: N000

GENERAL INSTRUCTIONS: Use to report current NBC situation and chemical units' activity. Send on a scheduled time line IAW unit SOPs, such as 0500, 1100, 1700, and 2300 daily. This is considered routine and is a normal report not indicated in NBC 1 through NBC 6 reports. Reference: FM 3-7.

LINE 1 – DATE AND TIME ______________________________ (DTG)

LINE 2 – UNIT ______________________________________ (Unit Making Report)

LINE 3 – THREATCON _________________________________ (Threatcon Level of Unit)

LINE 4 – MOPP _____________________________________ (Minimum MOPP Level of Unit)

LINE 5 – RADIATION STAT ____________________________ (Unit Radiation Status)

LINE 6 – ENEMY ______________________________________ (Enemy NBC Activity (Use, Nonuse, Any Enemy Chemical Units Detected))

LINE 7 – NBC UNITS __________________________________ (Status of Chemical Units or Elements (Location, Activity, Recon Status, Readiness Status))

a. No. of Smoke systems auth._____No. Opnl_____Projected Opnl in 24 Hrs._____
b. No. of Decon systems auth_____ No. Opnl_____Projected Opnl in 24 Hrs._____
c. No. of Recon systems auth_____ No. Opnl_____Projected Opnl in 24 Hrs._____

LINE 8 – NBC RECON _________________________________ (Reconnaissance Missions Conducted During Reporting Period)

Missions planned_____ DTG of exec._____Location _________ Unit supported_____Mission Purpose____________

LINE 9 – DECON ______________________________________ (Decon Missions Conducted)

Missions planned_____ DTG of exec._____Location _________ Unit supported_____Mission Purpose____________

LINE 10 – SMOKE _____________________________________ (Smoke Missions Conducted)

Missions planned_____ DTG of exec._____Location _________ Unit supported_____Mission Purpose____________

**Continued on next page

TITLE: NBC SITUATION REPORT [NBCSITREP] (CONTINUED)
REPORT NUMBER: N000

LINE 11 – NARRATIVE ______________________________ (Free Text for Additional Information Required for Clarification of Report)

LINE 12 – AUTHENTICATION __________________________ (Report Authentication)

TITLE: NBC 1/ROTA REPORT [NBC1]
REPORT NUMBER: N001 {USMTF # C488}

GENERAL INSTRUCTIONS: Use to provide the observer's initial report giving basic data on a nuclear attack, biological attack, chemical attack, or releases other than attack (ROTA). This NBC1/ROTA voice message accommodates all NBC1/ROTA reporting. Reference: ATP-45 and FM 3-7.

LINE 1 – DATE AND TIME ______________________________ (DTG)

LINE 2 – UNIT ______________________________ (Unit Making Report)

LINE 3 – EVENT ______________________________ (Type of Incident: NUCLEAR, NUC.NAV, BIOLOGICAL, BIO.NAV, CHEMICAL, or CHEM.NAV)

LINE 4 – ALFA ______________________________ (NBC Strike Serial Number)

LINE 5 – BRAVO ______________________________ (Position of Observer: UTM or Six-Digit Grid Coordinate With MGRS Grid Zone Designator)

LINE 6 – CHARLIE ______________________________ (Direction of Attack in Mils or Degrees as Measured Clockwise From Grid, True, or Magnetic North of Attack or ROTA From the Observer)

LINE 7 – DELTA ______________________________ (DTG of Detonation or Beginning of Attack or ROTA)

LINE 8 – ECHO ______________________________ (End DTG of Attack or ROTA)

LINE 9 – FOXTROT ______________________________ (UTM or Six-Digit Grid Coordinate With MGRS Grid Zone Designator of Attack and Code Used to Represent if Reported Location of Attack or ROTA is Actual or Estimated)

LINE 10 – GOLF ______________________________ (Vehicle or Device by Which Weapon Was Delivered or ROTA Was Released)

LINE 11 – HOTEL ______________________________ (Type of Burst, Biological/ Chemical Agent, and Persistency)

**Continued on next page.

TITLE: NBC 1/ROTA REPORT [NBC1] (CONTINUED)
REPORT NUMBER: N001 {USMTF # C488}

LINE 12 – INDIA ______	(Number of Munitions or Aircraft)
LINE 13 – JULIET ______	(Time in Seconds Denoting Flash-to-Bang DTG of Nuclear Attack)
LINE 14 – KILO ______	(Terrain and Vegetation Description or Crater Indicator and Width in Feet or Meters)
LINE 15 – LIMA ______	(Nuclear Burst Angular Cloud Width Measured at Five Minutes After Detonation Measured in Degrees or Mils)
LINE 16 – MIKE ______	(Stabilized Cloud Top or Bottom Height in Feet or Meters or Angular Cloud Top or Bottom Angle in Degrees or Mils Recorded at H+10 Minutes after Detonation)
LINE 17 – PAPA ALFA ROMEO ______	(Coordinate(s) of Hazardous Cloud or Area)
LINE 18 – PAPA BRAVO ROMEO ______	(Downwind Direction of Radioactive Cloud)
LINE 19 – SIERRA ______	(DTG of the Reading, for Nuclear Reporting, or Initial Detection Time, for Biological/ Chemical or ROTA Report)
LINE 20 – YANKEE ______	(Representative Downwind Direction and Wind Speed or Radial Line Sector)
LINE 21 – ZULU ALFA STABILITY ______	(Air Stability Indicator, Either Detailed or Simplified)
LINE 22 – ZULU ALFA TEMPERATURE ______	(Surface Air Temperature)
LINE 23 – ZULU ALFA HUMIDITY ______	(Relative Humidity Range)

**Continued on next page.

TITLE: NBC 1/ROTA REPORT [NBC1] (CONTINUED)
REPORT NUMBER: N001 {USMTF # C488}

LINE 24 – ZULU ALFA WEATHER ____________________ (Significant Weather Phenomena)

LINE 25 – ZULU ALFA COVER ____________________ (Cloud Cover)

LINE 26 – ZULU BRAVO ILLUMINATION____________________ (Illumination Time)

LINE 27 – TIME____________________ (DTG of Observation)

LINE 28 – NARRATIVE ____________________ (Free Text for Additional Information Required for Clarification of Report)

LINE 29 – AUTHENTICATION ____________________ (Report Authentication)

TITLE: NBC 2/ROTA REPORT [NBC2]
REPORT NUMBER: N002 {USMTF # G489}

GENERAL INSTRUCTIONS: Use to disseminate evaluated data of a nuclear attack, biological attack, chemical attack, or releases other than attack (ROTA). This NBC2/ROTA accommodates NBC2/ROTA (NUC, BIO, or CHEM) reporting requirements. Reference: ATP-45 (A) and FM 3-7.

LINE 1 – DATE AND TIME ______________________ (DTG)

LINE 2 – UNIT ______________________ (Unit Making Report)

LINE 3 – EVENT ______________________ (Type of Incident: NUCLEAR, NUC.NAV, BIOLOGICAL, BIO.NAV, CHEMICAL, or CHEM.NAV)

LINE 4 – ALFA ______________________ (NBC Strike Serial Number)

LINE 5 – DELTA ______________________ (DTG of Detonation or Beginning of Attack or ROTA)

LINE 6 – ECHO ______________________ (Biological or Chemical Events Only: End DTG of Attack or ROTA)

LINE 7 – FOXTROT ______________________ (UTM or Six-Digit Grid Coordinate With MGRS Grid Zone Designator of Attack and Code Used to Represent if Reported Location of Attack or ROTA is Actual or Estimated)

LINE 8 – GOLF ______________________ (Vehicle or Device by Which Weapon Was Delivered or ROTA Was Released)

LINE 9 – HOTEL ______________________ (Type of Burst, Biological/ Chemical Agent, and Persistency)

LINE 10 – INDIA ______________________ (Number of Munitions or Aircraft)

LINE 11 – KILO ______________________ (Terrain and Vegetation Description or Crater Indicator and Width in Feet or Meters)

**Continued on next page.

TITLE: NBC 2/ROTA REPORT [NBC2] (CONTINUED)
REPORT NUMBER: N002 {USMTF # G489}

LINE 12 – NOVEMBER______________	(Estimated Nuclear Yield in Kilotons Released From Detonation of the Nuclear Weapon)
LINE 13 – YANKEE______________	(Radial Line Sector or Representative Downwind Direction and Wind Speed)
LINE 14 – ZULU ALFA STABILITY ______________	(Air Stability Indicator, Either Detailed or Simplified)
LINE 15 – ZULU ALFA TEMPERATURE______________	(Surface Air Temperature)
LINE 16 – ZULU ALFA HUMIDITY ______________	(Relative Humidity Range Indicator)
LINE 17 – ZULU ALFA WEATHER ______________	(Significant Weather Phenomena)
LINE 18 – ZULU ALFA COVER ______________	(Cloud Cover)
LINE 19 – TIME______________	(DTG of Observation)
LINE 20 – NARRATIVE ______________	(Free Text for Additional Information Required for Clarification of Report)
LINE 21 – AUTHENTICATION ______________	(Report Authentication)

TITLE: NBC 3/ROTA REPORT [NBC3]
REPORT NUMBER: N003 {USMTF # C443}

GENERAL INSTRUCTIONS: Use to pass immediate warning of predicted contamination and hazard areas following NBC attacks or releases other than attack (ROTA) hazards. This NBC3/ROTA voice message accommodates all NBC3/ROTA (NUC, BIO, or CHEM) reporting requirements. Reference: ATP-45 and FM 3-7.

LINE 1 – DATE AND TIME ____________	(DTG)
LINE 2 – UNIT ____________	(Unit Making Report)
LINE 3 – EVENT ____________	(Type of Incident: NUCLEAR, NUC.NAV, BIOLOGICAL, BIO.NAV, CHEMICAL, or CHEM.NAV)
LINE 4 – ALFA ____________	(NBC Strike Serial Number)
LINE 5 – DELTA ____________	(DTG of Detonation or Beginning of Attack or ROTA)
LINE 6 – FOXTROT ____________	(Location of Attack or Event, Area Attacked/Affected)
LINE 7 – GOLF ____________	(Means of Delivery/ROTA Classification)
LINE 8 – HOTEL ____________	(Type Burst/Type Agent, Persistency/TM, UNID Number)
LINE 9 – NOVEMBER ____________	(Estimated Yield)
LINE 10 – PAPA ALFA ____________	(UTM or Six-Digit Grid Coordinate With MGRS Grid Zone Designator of Predicted Hazardous Cloud or Area)
LINE 11 – PAPA BRAVO ____________	(Duration of Hazard in Days or Fraction of Days)
LINE 12 – YANKEE ____________	(Radial Line Sector or the Representative Downwind Direction and Wind Speed)

**Continued on next page.

TITLE: NBC 3/ROTA REPORT [NBC3] (CONTINUED)
REPORT NUMBER: N003 {USMTF # C443}

LINE 13 – ZULU	______	(Effective Wind Speed in KM per Hour or Knots, Downwind Distance of Zone 1 in KM or Nautical Miles, and Cloud Radius in KM or Nautical Miles)
LINE 14 – ZULU ALFA STABILITY	______	(Air Stability Indicator, Either Detailed or Simplified)
LINE 15 – ZULU ALFA TEMPERATURE	______	(Surface Air Temperature)
LINE 16 – ZULU ALFA HUMIDITY	______	(Relative Humidity Range Indicator)
LINE 17 – ZULU ALFA WEATHER	______	(Significant Weather Phenomena)
LINE 18 – ZULU ALFA COVER	______	(Cloud Cover)
LINE 19 – ZULU BRAVO	______	(Remarks (CHEM))
LINE 20 – TIME	______	(DTG of Observation)
LINE 21 – NARRATIVE	______	(Free Text for Additional Information Required for Clarification of Report)
LINE 22 – AUTHENTICATION	______	(Report Authentication)

TITLE: NBC 4 REPORT [NBC4]
REPORT NUMBER: N004 {USMTF # C447}

GENERAL INSTRUCTIONS: Use to report NBC monitoring and survey results. This NBC4 voice message accommodates all NBC4 (NUC, BIO, or CHEM) reporting. Reference: FM 3-7.

LINE 1 – DATE AND TIME ______________________ (DTG)

LINE 2 – UNIT ______________________ (Unit Making Report)

LINE 3 – EVENT ______________________ (Type of Incident: NUCLEAR, NUC.NAV, BIOLOGICAL, BIO.NAV, CHEMICAL, or CHEM.NAV)

LINE 4 – ALFA ______________________ (NBC Strike Serial Number)

LINE 5 – HOTEL ______________________ (Type of Burst, Biological/ Chemical Agent, and Persistency)

LINE 6 – KILO ______________________ (Terrain and Vegetation Description or Crater Indicator and Width in Feet or Meters)

LINE 7 – QUEBEC ______________________ (UTM or Six-Digit Grid Coordinate With MGRS Grid Zone Designator of the Reading or Sample and the Categories/ Type of Sample)

LINE 8 – ROMEO ______________________ (Level of Radiation, Measured in cGy/H, Dose Rate Trend, and Relative or Actual Radiation Decay Rate)

LINE 9 – SIERRA ______________________ (DTG the Contamination Was Detected or DTG of Reading or Sample)

LINE 10 – TIME ______________________ (DTG of Observation)

LINE 11 – NARRATIVE ______________________ (Free Text for Additional Information Required for Clarification of Report)

LINE 12 – AUTHENTICATION ______________________ (Report Authentication)

TITLE: NBC 5 REPORT [NBC5]
REPORT NUMBER: N005 {USMTF # C501}

GENERAL INSTRUCTIONS: Use to pass information on areas of actual nuclear, biological, or chemical contamination. This NBC5 voice message accommodates all NBC5 (NUC, BIO, or CHEM) reporting requirements by selected use of sets. Reference: FM 3-7.

LINE 1 – DATE AND TIME ________________	(DTG)
LINE 2 – UNIT ________________	(Unit Making Report)
LINE 3 – EVENT ________________	(Type of Incident: NUCLEAR, BIOLOGICAL, or CHEMICAL)
LINE 4 – ALFA ________________	(NBC Strike Serial Number)
LINE 5 – DELTA ________________	(DTG of Detonation or Beginning of Attack)
LINE 6 – FOXTROT ________________	(UTM or Six-Digit Grid Coordinate With MGRS Grid Zone Designator of Attack and Code Used to Represent if Reported Location of Attack is Actual or Estimated)
LINE 7 – HOTEL ________________	(Type of Burst, Biological/ Chemical Agent, and Persistency)
LINE 8 – OSCAR ________________	(Reference: DTG for Estimated Contours When Not H+1 Hour)
LINE 9 – SIERRA ________________	(DTG Contamination Was Detected or DTG of Reading)
LINE 10 – TANGO ________________	(H+1 DTG of Reading)
LINE 11 – UNIFORM ________________	(Coordinates of 1,000 cGy/H Contour Line-Red)
LINE 12 – VICTOR ________________	(Coordinates of 300 cGy/H Contour Line-Green)
LINE 13 – WHISKEY ________________	(Coordinates of 100 cGy/H Contour Line-Blue)

**Continued on next page.

TITLE: NBC 5 REPORT [NBC5] (CONTINUED)
REPORT NUMBER: N005 {USMTF # C501}

LINE 14 – XRAY	______________________________	(Coordinates of 20 cGy/H Contour Line or Area of Biological Chemical Contamination - Black or Yellow)
LINE 15 – TIME	______________________________	(DTG of Observation)
LINE 16 – NARRATIVE	______________________________	(Free Text for Additional Information Required for Clarification of Report)
LINE 17 – AUTHENTICATION	______________________________	(Report Authentication)

TITLE: NBC 6/ROTA REPORT [NBC6]
REPORT NUMBER: N006 {USMTF # C506}

GENERAL INSTRUCTIONS: Use to pass detailed information on chemical and biological attacks and releases other than attack (ROTA) hazards. This NBC6/ROTA voice message accommodates all NBC6/ROTA (BIO or CHEM) reporting requirements. Reference: ATP-45 (A) and FM 3-7.

LINE 1 – DATE AND TIME ______________________	(DTG)
LINE 2 – UNIT ______________________	(Unit Making Report)
LINE 3 – EVENT ______________________	(Type of Incident: BIOLOGICAL or CHEMICAL)
LINE 4 – ALFA ______________________	(NBC Strike Serial Number)
LINE 5 – DELTA ______________________	(DTG of Detonation or Beginning of Attack or Release)
LINE 6 – ECHO ______________________	(Day-Time Attack or Release Ended)
LINE 7 – FOXTROT ______________________	(UTM or Six-Digit Grid Coordinate With MGRS Grid Zone Designator of Attack and Code Used to Represent if Reported Location of Attack or ROTA Release is Actual or Estimated)
LINE 8 – GOLF ______________________	(Means by Which Chemical/ Biological Agent Weapon or ROTA Was Delivered or Released)
LINE 9 – HOTEL ______________________	(Type of Burst, Biological/ Chemical Agent, and Persistency)
LINE 10 – INDIA ______________________	(Number of Munitions or Aircraft)
LINE 11 – KILO ______________________	(Description of Terrain and Vegetation)
LINE 12 – MIKE ______________________	(Enemy Action Before and After Attack and Effect on Troops)

**Continued on next page

TITLE: NBC 6/ROTA REPORT [NBC6] (CONTINUED)
REPORT NUMBER: N006 {USMTF # C506}

LINE 13 – QUEBEC ______	(Location and Type of Sample)
LINE 14 – SIERRA ______	(DTG Contamination Detected)
LINE 15 – TANGO ______	(DTG of Latest Survey)
LINE 16 – XRAY ______	(Area of Actual Contamination)
LINE 17 – YANKEE ______	(Downwind Direction and Wind Speed)
LINE 18 – ZULU ALFA ______	(Actual Weather Conditions)
LINE 19 – TIME ______	(DTG of Observation)
LINE 20 – NARRATIVE ______	(Free Text for Additional Information Required for Clarification of Report)
LINE 20L1 POSITION OF OBSERVER ______	(UTM or Six-Digit Grid Coordinate With MGRS Grid Zone Designator of Observer)
LINE 20A1 HELICOPTER TO AEROSOL RANGE ______	(Range in Meters From Helicopter to Aerosol)
LINE 20A2 AEROSOL WIDTH ______	(Aerosol Width)
LINE 20A3 AEROSOL HEIGHT ______	(Aerosol Height (optional))
LINE 20A4 AEROSOL HEIGHT ABOVE GROUND ______	(Aerosol Height in Meters AGL)
LINE 20Z DETECTION TIME ______	(DTG the Aerosol Was Detected)
LINE 20E1 HELICOPTER ALTITUDE ______	(Helicopter Altitude)
LINE 20E2 HELICOPTER HEADING ______	(Helicopter Heading in Degrees)

**Continued on next page

TITLE: NBC 6/ROTA REPORT [NBC6] (CONTINUED)
REPORT NUMBER: N006 {USMTF # C506}

LINE 20E3 HELICOPTER CRAB ______________________________ (Helicopter Crab Angle)

LINE 20R1 TYPE OF REPORT________________________________ (L, F, LC)

LINE 20R2 FLIGHT TIME REMAINING______________________ (Flight Time Remaining)

LINE 20R3 GROUND TRACK ________________________________ (Ground Track)

LINE 21 – AUTHENTICATION ______________________________ (Report Authentication)

**Add LINES 20 L1-R3 for the Long Range Biological Standoff Detection System (LRBSDS).

TITLE: OPERATION ORDER [ORDER]
REPORT NUMBER: O001

GENERAL INSTRUCTIONS: Use to transmit the standard five-paragraph operation order (OPORD) and operation plan (OPLAN) that provide plans, instructions, and directives to subordinate and supporting military organizations. Reference: FM 101-5, FM 71-100, and FM 100-15.

LINE 1 – DATE AND TIME ______________________ (DTG)

LINE 2 – UNIT ______________________ (Unit Making Report)

LINE 3 – OPERATION (PLAN) ORDER ______________________ (Order Number)

LINE 4 – REFERENCES ______________________ (References)

LINE 5 – TIME ZONE ______________________ (TZ Used in OPORD/OPLAN)

LINE 6 – TASK ORGANIZATION ______________________ (TO of Unit)

LINE 7 – SITUATION

a. ENEMY ______________________ (Enemy Forces)

b. FRIENDLY ______________________ (Friendly Forces)

c. ATTACH/DET ______________________ (Attachments/Detachments)

d. ASSUMPTIONS (OPLAN ONLY) ______________________ (OPLAN only)

LINE 8 – MISSION ______________________

LINE 9 – EXECUTION INTENT ______________________

a. CONCEPT ______________________ (Concept of Operation)

(1) MANEUVER ______________________ (Maneuver Units)

(2) FIRES ______________________ (Fire Support Elements)

(3) RECON AND SURVEILLANCE ______________________ (R&S PLAN)

(4) INTELLIGENCE ______________________ (MI Elements)

(5) ENGINEER ______________________ (Engineer Elements)

(6) AIR DEFENSE ______________________ (ADA Elements)

(7) INFORMATION OPERATIONS ______________________ (Information Ops Elements)

**Continued on next page.

TITLE: OPERATION ORDER [ORDER] (CONTINUED)
REPORT NUMBER: O001

b. TASKS TO MANEUVER UNITS ________________ ________________ ________________	(Tasks to Maneuver Units)
c. TASKS TO COMBAT SUPPORT UNITS__________	(Tasks to CS Units)
(1) INTELLIGENCE________________	(Tasks to MI Units)
(2) ENGINEER ________________	(Tasks to Engineer)
(3) FIRES ________________	(Tasks to Fire Support Elements)
(4) AIR DEFENSE ________________	(Tasks to ADA Elements)
(5) SIGNAL________________	(Tasks to Signal Elements)
(6) NBC________________	(Tasks to NBC Elements)
(7) PROVOST MARSHAL________________	(Tasks to PM Elements)
(8) PSYOP________________	(Tasks to Special Ops Elements)
(9) CIVIL MILITARY ________________	(Tasks to CA Elements)
(10) AS REQUIRED________________	
d. COORDINATING INSTRUCTIONS ____________	(Coordinating Instructions)
(1) TIME ORDER/PLAN IN EFFECT____________	(DTG of Effective Time of Plan/Order)
(2) CCIR ________________	(CDR's CCIR by Priority and Phase)
(3) RISK MGMT________________	(Risk Reduction Control Measures)
(4) ROE________________	(Rules of Engagement)
(5) EPA ________________	(Environmental Concerns)
(6) FORCE PROTECTION ________________	(Force Protection Measures)
(7) AS REQUIRED________________	(When Needed)

**Continued on next page.

TITLE: OPERATION ORDER [ORDER] (CONTINUED)
REPORT NUMBER: O001

LINE 10 – SERVICE AND SUPPORT

a. SUPPORT CONCEPT ______ (Concept of Support)

b. MATERIAL AND SERVICES ______

c. MEDICAL EVAC AND HOSPITALIZATION ______

d. PERSONNEL ______

e. CIVIL MILITARY ______

f. AS REQUIRED ______

LINE 11 – COMMAND AND SIGNAL

a. COMMAND ______ (Command)

b. SIGNAL ______ (Signal)

LINE 12 – ACKNOWLEDGE ______ (Mandatory)

LINE 13 – CDR NAME/RANK ______ (Commander's Name and Rank)

LINE 14 – ANNEX A ______ (TO Instructions and Format)

LINE 15 – ANNEX B ______ (INTEL Instructions and Format)

LINE 16 – ANNEX C ______ (OPS Overlay Instructions and Format)

LINE 17 – ANNEX D ______ (Fire Support Instructions and Format)

LINE 18 – ANNEX E ______ (ROE Instructions and Format)

LINE 19 – ANNEX F ______ (Engineer Instructions and Format)

LINE 20 – ANNEX G ______ (ADA Instructions and Format)

LINE 21 – ANNEX H ______ (Signal Instructions and Format)

LINE 22 – ANNEX I ______ (CSS Instructions and Format)

**Continued on next page.

TITLE: OPERATION ORDER [ORDER] (CONTINUED)
REPORT NUMBER: O001

LINE 23 – ANNEX J ________	(NBC Ops Instructions and Format)
LINE 24 – ANNEX K ________	(PM Instructions and Format)
LINE 25 – ANNEX L ________	(R&S Instructions and Format)
LINE 26 – ANNEX M ________	(Deep Ops Instructions and Format)
LINE 27 – ANNEX N ________	(Rear Ops Instructions and Format)
LINE 28 – ANNEX O ________	(A2C2 Instructions and Format)
LINE 29 – ANNEX P ________	(CCW Instructions and Format)
LINE 30 – ANNEX Q ________	(OPSEC Instructions and Format)
LINE 31 – ANNEX R ________	(PSYOP Instructions and Format)
LINE 32 – ANNEX S ________	(Deception Instructions and Format)
LINE 33 – ANNEX T ________	(EW Instructions and Format)
LINE 34 – ANNEX U ________	(Civil-Military OPS Instructions and Format)
LINE 35 – ANNEX V ________	(PA Instructions and Format)
LINE 36 – NARRATIVE ________	(Free Text for Additional Information Required for Clarification of Report)
LINE 37 – AUTHENTICATION ________	(Report Authentication)

TITLE: OPERATION REPORT [OPREP]
REPORT NUMBER: O005

GENERAL INSTRUCTIONS: Use to report operational situations, problems, recommended or intended courses of action, and other items not reported elsewhere. Send as determined by higher headquarters or by necessity of subordinate headquarters. Reference: FM 100-15, FM 71-100, and FM 101-5.

LINE 1 – DATE AND TIME ____________________ (DTG)

LINE 2 – UNIT ____________________ (Unit Making Report)

LINE 3 – COMMAND POST ____________________ (CP/LZ Locations for TAC CP, Main CP, and Rear CP)

LINE 4 – FLOT ____________________ (FLOT at Least Three Grids)

LINE 5 – FEBA ____________________ (FEBA at Least Three Grids)

LINE 6 – OPSUM ____________________ (Brief Summary of Reporting Unit's Activity and Task Organization)

LINE 7 – ENEMY ____________________ (Enemy Activity in Reporting Unit's Area of Interest and Operation)

LINE 8 – UNIT STAT ____________________ (Unit Statistics)

PERSONNEL ____________________

EQUIPMENT ____________________

CLASS OF SUPPLY ____________________

LINE 9 – EVALUATION ____________________ (Reporting Commander's Overall Evaluation to Include Mission, Commander's Overall Evaluation of Reporting Unit's Ability to Accomplish its Mission.)

LINE 10 – NARRATIVE ____________________ (Free Text for Additional Information Required for Clarification of Report)

LINE 11 – AUTHENTICATION ____________________ (Report Authentication)

TITLE: OPERATIONS PLAN CHANGE [PLANORDCHG]
REPORT NUMBER: O010 {USMTF # E400}

GENERAL INSTRUCTIONS: Use to update or change existing operations plans. (Note: Use FRAGO to change OPORD.) Reference: FM 100-15, FM 71-100, and FM 101-5.

LINE 1 – DATE AND TIME ______________________	(DTG)
LINE 2 – UNIT ______________________	(Unit Making Report)
LINE 3 – ORIGINATOR ______________________	(Military Designation of Responsible Authority of the Operations Plan)
LINE 4 – NUMBER ______________________	(Operations Plan Number)
LINE 5 – ANNEX ______________________	(Annex Number Assigned to the Specific Operation Plan Being Changed)
LINE 6 – APPENDIX ______________________	(Appendix Number Assigned to the Specific Operation Plan Being Changed)
LINE 7 – TAB ______________________	(Tab Number Assigned to the Specific Operation Plan Being Changed)
LINE 8 – ENCLOSURE ______________________	(Enclosure Number Assigned to the Specific Operation Plan Being Changed)
LINE 9 – PAGE ______________________	(Page Number Assigned to the Specific Operation Plan Being Changed)
LINE 10 – PARAGRAPH ______________________	(Paragraph Number Assigned to the Specific Operation Plan Being Changed)
LINE 11 – LINE ______________________	(Line Number Assigned to the Specific Operation Plan Being Changed)
LINE 12 – NARRATIVE ______________________	(Free Text for Additional Information Required for Clarification of Report)
LINE 13 – AUTHENTICATION ______________________	(Report Authentication)

TITLE: OPERATIONS SUMMARY [OPSUM]
REPORT NUMBER: O015 {USMTF # G423}

GENERAL INSTRUCTIONS: Use to provide a commander's summary of significant operations to the next higher commander. Reference: FM 100-15, FM 71-100, and FM 101-5.

LINE 1 – DATE AND TIME ______________________________ (DTG)

LINE 2 – UNIT ______________________________ (Unit Making Report)

LINE 3 – AIR ______________________________ (Number of Sorties Tasked, Flown, Successful, Kills Claimed)

LINE 4 – GROUND ______________________________ (Ground Operations Summarized)

LINE 5 – MARITIME______________________________ (Maritime Operations Summarized)

LINE 6 – NARRATIVE ______________________________ (Free Text for Additional Information Required for Clarification of Report)

LINE 7 – AUTHENTICATION ______________________________ (Report Authentication)

TITLE: PATROL REPORT [PATROLREP]
REPORT NUMBER: P001

GENERAL INSTRUCTIONS: Use to report information and combat actions obtained on mounted and dismounted patrols. Reference: FM 17-98, FM 17-95, FM 7-92, and FM 34-3.

LINE 1 – DATE AND TIME ______________	(DTG)
LINE 2 – UNIT ______________	(Unit Making Report)
LINE 3 – DPT & RTN DTG ______________	(Patrol Departure and Return DTG)
LINE 4 – ROUTE ______________	(Patrol Route (if Applicable))
LINE 5 – CHECK POINTS ______________	(Check Points Patrol Route Will Follow)
LINE 6 – RESULTS ______________	(Results (if Specified in Mission Task and Purpose))
LINE 7 – CONDITION OF PATROL ______________	(Status Report of Patrol, Friendly Losses)
LINE 8 – SALUTE ______________	(Enemy Encountered, SALUTE)
LINE 9 – BDA ENEMY ______________	(Battle Damage Assessment Enemy)
LINE 10 – ROUTE RECON RPT ______________	(Description of Terrain, Route Recon Report, Correction to Maps (if Applicable))
LINE 11 – NARRATIVE ______________	(Free Text for Additional Information Required for Clarification of Report)
LINE 12 – AUTHENTICATION ______________	(Report Authentication)

TITLE: PERSONNEL STATUS REPORT [PERSTAT]
REPORT NUMBER: P005 {USMTF # G880}

GENERAL INSTRUCTIONS: Use to report the status of the unit's personnel. Reference: FM 12-6.

LINE 1 – DATE AND TIME	______________	(DTG)
LINE 2 – UNIT	______________	(Unit Making Report)
LINE 3 – FROM	______________	(DTG for the Beginning of Period Applying to Personnel Information)
LINE 4 – TO	______________	(DTG for the End of Report Period)
LINE 5 – UNIT	______________	(Designation of the Unit for Which the Personnel Status Information is Being Submitted)
LINE 6 – AUTHORIZED	______________	(Number of Personnel Authorized by Personnel Classification)
LINE 7 – ASSIGNED	______________	(Number of Personnel Assigned by Personnel Classification)
LINE 8 – ON HAND	______________	(Number of Personnel on Hand by Personnel Classification)
LINE 9 – GAINS	______________	(Number of Personnel Gains by Personnel Classification)
LINE 10 – REPLACEMENTS	______________	(Number of Personnel Gained That Are Replacements by Personnel Classification)
LINE 11 – RETURNED TO DUTY	______________	(Number of Personnel Gained That Have Been Returned to Duty Through Medical Channels by Personnel Classification)
LINE 12 – KILLED	______________	(Number of KIA by Personnel Classification)
LINE 13 – WOUNDED	______________	(Number of WIA by Personnel Classification)

**Continued on next page.

TITLE: PERSONNEL STATUS REPORT [PERSTAT] (CONTINUED)
REPORT NUMBER: P005 {USMTF # G880}

LINE 14 – NONBATTLE LOSS ________________	(Number of Disease Nonbattle Injury Loss by Personnel Classification)
LINE 15 – MISSING ________________	(Number of MIA by Personnel Classification)
LINE 16 – DESERTERS ________________	(Number of Deserters by Personnel Classification)
LINE 17 – AWOL________________	(Number Absent Without Leave by Personnel Classification)
LINE 18 – CAPTURED ________________	(Number of Enemy Personnel Captured During the Reporting Period)

**Repeat lines 5 through 18 to indicate the personnel summary of additional units. Assign sequential line numbers to succeeding iterations; for example, first iteration 5 through 18; second iteration 5a through 18a; third iteration 5b through 18b, and so on.

LINE 19 – NARRATIVE ________________	(Free Text for Additional Information Required for Clarification of Report)
LINE 20 – AUTHENTICATION ________________	(Report Authentication)

TITLE: PRELIMINARY TECHNICAL REPORT [PRETECHREP]
REPORT NUMBER: P010 {USMTF # F105}

GENERAL INSTRUCTIONS: Use to disseminate initial technical information on captured or acquired foreign material and ordnance. Reference: FM 34-54 and FM 34-60.

LINE 1 – DATE AND TIME ________________________________ (DTG)

LINE 2 – UNIT ________________________________ (Unit Making Report)

LINE 3 – MATERIEL CATEGORY ________________________________ (Category of Materiel the Acquired Items Belong to)

LINE 4 – UNIT ID ________________________________ (Identification of the Unit Acquiring the Materiel)

LINE 5 – DAY TIME ACQUIRED ________________________________ (DTG When the Materiel Was Acquired)

LINE 6 – LOCATION ACQUIRED________________________________ (UTM or Six-Digit Grid Coordinate With MGRS Grid Zone Designator Where Acquired)

LINE 7 – CIRCUMSTANCES ________________________________ (Brief Description of the Circumstances Surrounding Acquisition of the Materiel)

LINE 8 – CONTACT NAME________________________________ (Name of Reporting Unit POC)

LINE 9 – RANK OR RATE ________________________________ (Rank or Rate/Rating of Reporting Unit POC)

LINE 10 – PRIMARY PHONE ________________________________ (Primary Telephone Number of Reporting Unit POC)

LINE 11 – PRIMARY FREQUENCY ________________________________ (Primary Radio Frequency of Reporting Unit Using MHz, Frequency Designator, or Frequency)

LINE 12 – POC LOCATION ________________________________ (POC Location Using UTM or Six-Digit Grid Coordinate with MGRS Grid Zone Designator)

LINE 13 – SECONDARY PHONE ________________________________ (Secondary Telephone Number of Reporting Unit POC)

**Continued on next page.

TITLE: PRELIMINARY TECHNICAL REPORT [PRETECHREP] (CONTINUED)
REPORT NUMBER: P010 {USMTF # F105}

LINE 14 – SECONDARY FREQUENCY ______________________ (Secondary Radio Frequency of Reporting Unit Using MHz, Frequency Designator, or Frequency)

LINE 15 – OTHER INFORMATION ________________________ (Other Additional Information Concerning Unexploded Ordnance Such as Tentative Render Safe Procedures, Other Descriptions)

**Repeat lines 3 through 15 to report multiple materiel acquisitions. Assign sequential line numbers to succeeding iterations; for example, first iteration 3 through 15; second iteration 3a through 15a; third iteration 3b through 15b; and so on.

LINE 16 – NARRATIVE ________________________________ (Free Text for Additional Information Required for Clarification of Report)

LINE 17 – AUTHENTICATION ____________________________ (Report Authentication)

TITLE: PSYCHOLOGICAL OPERATIONS REPORT [PSYOPREP]
REPORT NUMBER: P015 {USMTF # C500}

GENERAL INSTRUCTIONS: Use to provide information on current and projected psychological operations activities. Reference: FM 33-1.

LINE 1 – DATE AND TIME	______________	(DTG)
LINE 2 – UNIT	______________	(Unit Making Report)
LINE 3 – ENVIRONMENT	______________	(Describe the General PSYOP Environment, to Include Friendly PSYOP Effectiveness, and Hostile PSYOP Assets or Products Encountered)
LINE 4 – ACTIVITIES	______________	(Enter the Disposition of PSYOP Forces and Operations)
LINE 5 – PROJECTED ACTIVITIES	______________	(Enter the Projected Disposition of Forces Next 24 Hours and Projected Operations)
LINE 6 – LOGISTICS	______________	(Enter the Logistics Situation and Requirements)
LINE 7 – EVALUATION	______________	(Enter the Commander's Evaluation)
LINE 8 – NARRATIVE	______________	(Free Text for Additional Information Required for Clarification of Report)
LINE 9 – AUTHENTICATION	______________	(Report Authentication)

TITLE: PUBLIC AFFAIRS OPERATION REPORT [PUBAFFOPSREP]
REPORT NUMBER: P020

GENERAL INSTRUCTIONS: Use to convey status of PA operations by PAO. Reference: FM 46-1.

LINE 1 – DATE AND TIME ________________________________ (DTG)

LINE 2 – UNIT ________________________________ (Unit Making Report)

LINE 3 – SUMMARY ________________________________ (Summary of PAO Activities)

a. Newspaper Production/Distribution Schedule____________

b. Number and Type of Media Escorted by Public Affairs

c. Personnel in the last 24 hours ________________________

d. Radio Broadcast Schedule___________________________

e. Radio Broadcast Wattage ___________________________

LINE 4 – ASSESSMENT ________________________________ (PAO Assessment)

a. Field Newspaper Produced on Schedule ________________ (YES or NO)

b. Number of MOS-Qualified Journalist Authorized _________ (NOH)

c. Number of Digital Cameras (Still) Authorized ___________ (NOH – OR)

d. Number of Military Journalists in Sector _______________

e. Number of Video Cameras in Sector___________________ (Number Operational)

LINE 5 – NARRATIVE ________________________________ (Free Text for Additional Information Required for Clarification of Report)

LINE 6 – AUTHENTICATION ___________________________ (Report Authentication)

TITLE: RADAR STATUS REPORT (RADSTAT)
REPORT NUMBER: R001

GENERAL INSTRUCTIONS: Use to report radar's unit section number, type, location, and azimuth of search in the counterfire headquarters. If the radar is damaged, destroyed, or in maintenance, add that information in LINE 7 (Narrative). Reference: FM 6-20-10 and FM 6-121.

LINE 1 – DATE AND TIME ______________________________ (DTG)

LINE 2 – UNIT ______________________________ (Unit Making Report)

LINE 3 – SECTION ______________________________ (Section)

LINE 4 – RADAR TYPE ______________________________ (Type Radar (Q36, Q37))

LINE 5 – RADAR LOCATION ______________________________ (Radar Grid)

LINE 6 – AZ ______________________________ (Azimuth of Search)

LINE 7 – NARRATIVE ______________________________ (Free Text for Additional Information Required for Clarification of Report)

LINE 8 – AUTHENTICATION ______________________________ (Report Authentication)

TITLE: REAR AREA PROTECTION UNIT STATUS [RAPSTAT]
REPORT NUMBER: R005 {USMTF # B930}

GENERAL INSTRUCTIONS: Use to report the status of subordinate units available for RAP-type missions. Reference: FM 100-15 and FM 19-1.

LINE 1 – DATE AND TIME ______	(DTG)
LINE 2 – UNIT ______	(Unit Making Report)
LINE 3 – UNITS AND LOCATION ______	(Units That Are Available for RAP Missions and Their UTM or Six-Digit Grid Coordinate With MGRS Grid Zone Designator)
LINE 4 – PERSONNEL ______	(Personnel Available for RAP Missions by Personnel Class and Unit)
LINE 5 – EQUIPMENT ______	(Quantity and Type of Equipment and Weapons by Unit Available for RAP Missions)
LINE 6 – EMPLOYMENT LIMITATIONS ______	(Employment Limitations of the Reporting Unit)
LINE 7 – NARRATIVE ______	(Free Text for Additional Information Required for Clarification of Report)
LINE 8 – AUTHENTICATION ______	(Report Authentication)

TITLE: REAR AREA SECURITY ACTIVITIES [RASACT]
REPORT NUMBER: R010 {USMTF # D920}

GENERAL INSTRUCTIONS: Use to report rear area security activities. Reference: FM 100-15 and FM 19-1.

LINE 1 – DATE AND TIME ______________	(DTG)
LINE 2 – UNIT ______________	(Unit Making Report)
LINE 3 – SECURITY ACTIVITY ______________	(Type Security Activity That Will Occur)
LINE 4 – SECURITY UNITS ______________	(Number and Type Units That Will be Involved)
LINE 5 – FROM ______________	(Estimated DTG That Security Activity Will Start)
LINE 6 – TO ______________	(Estimated DTG That Security Activity Will End)
LINE 7 – LOCATION ______________	(UTM or Six-Digit Grid Coordinate With MGRS Grid Zone Designator Where Security Activity Will be Conducted)
LINE 8 – INITIAL POINT ______________	(Initial or Start Point of the Location UTM or Six-Digit Grid Coordinate With MGRS Grid Zone Designator)
LINE 9 – TRACE PLOT ______________	(Trace Plot Defined by a Series of Points in Required Sequence Using UTM or Six-Digit Grid Coordinate With MGRS Grid Zone Designator)
LINE 10 – CALL SIGN ______________	(Call Sign of Unit Coordinating the Security Activity)
LINE 11 – PRIMARY ______________	(Primary Radio Frequency of Unit Coordinating the Security Activity)

**Continued on next page.

TITLE: REAR AREA SECURITY ACTIVITIES [RASACT] (CONTINUED)
REPORT NUMBER: R010 {USMTF # D920}

LINE 12 – SECONDARY__________________________________ (Secondary Radio Frequency of Unit Coordinating the Security Activity)

LINE 13 – NARRATIVE __________________________________ (Free Text for Additional Information Required for Clarification of Report)

**Repeat lines 3 through 13 to identify multiple requests for security assistance. Assign sequential line numbers to succeeding iterations; for example, first iteration 3 through 13; second iteration 3a through 13a; third iteration 3b through 13b; and so on.

LINE 14 – AUTHENTICATION ______________________________ (Report Authentication)

TITLE: REAR AREA SECURITY REQUEST [RASREQ]
REPORT NUMBER: R015 {USMTF # D004}

GENERAL INSTRUCTIONS: Use to request rear area security assistance. Reference: FM 100-15, FM 71-100, and FM 19-1.

LINE 1 – DATE AND TIME ________________	(DTG)
LINE 2 – UNIT ________________	(Unit Making Report)
LINE 3 – REQUESTER ________________	(Identifier of Unit/Agency Requesting Security Assistance)
LINE 4 – SECURITY ACTIVITY ________________	(Security Activity Being Requested)
LINE 5 – UNITS ________________	(Number and Type Units Required to Conduct the Mission; if Unknown, State UNKNOWN)
LINE 6 – DATE AND TIME ________________	(Date and Time Assistance is Required)
LINE 7 – DURATION ________________	(Estimated Duration Time of the Mission in Hours)
LINE 8 – LOCATION ________________	(UTM or Six-Digit Grid Coordinate With MGRS Grid Zone Designator Where Security Assistance Mission Will be Conducted)
LINE 9 – INITIAL POINT ________________	(Initial or Start Point Location Using UTM or Six-Digit Grid Coordinate With MGRS Grid Zone Designator)
LINE 10 – TRACE PLOT ________________	(Trace Plot Defined by a Series of Points in Sequence UTM or Six-Digit Grid Coordinate With MGRS Grid Zone Designator)
LINE 11 – CALL SIGN ________________	(Call Sign of Unit Requesting the Security Activity)
LINE 12 – PRIMARY ________________	(Primary Radio Frequency of Unit Requesting the Security Activity)

**Continued on next page.

TITLE: REAR AREA SECURITY REQUEST [RASREQ] (CONTINUED)
REPORT NUMBER: R015 {USMTF # D400}

LINE 13 – SECONDARY__________________________________ (Secondary Radio Frequency of Unit Requesting the Security Activity)

LINE 14 – NARRATIVE __________________________________ (Free Text for Additional Information Required for Clarification of Report)

**Repeat lines 3 through 14 to identify multiple requests for security assistance. Assign sequential line numbers to succeeding iterations; for example, first iteration 3 through 14; second iteration 3a through 14a; third iteration, 3b through 14b; and so on.

LINE 15 – AUTHENTICATION ______________________________ (Report Authentication)

TITLE: RECONNAISSANCE EXPLOITATION REPORT [RECCEXREP]
REPORT NUMBER: R020 {USMTF # C101}

GENERAL INSTRUCTIONS: Use to provide an abbreviated imagery interpretation report in a tactical situation.

LINE 1 – DATE AND TIME ______________________________ (DTG)

LINE 2 – UNIT ______________________________ (Unit Making Report)

LINE 3 – SERIAL ______________________________ (Originator's Report Serial Number)

LINE 4 – REQUEST ______________________________ (Requestor Identity and Serial Number)

LINE 5 – PROJECT ______________________________ (Project Identifier Code)

LINE 6 – MISSION ______________________________ (Mission Number)

LINE 7 – DATE ______________________________ (Mission Date, DTG)

LINE 8 – ITEM ______________________________ (Item Number)

LINE 9 – NAME ______________________________ (Target Name or Description)

LINE 10 – NUMBER ______________________________ (BE Number)

LINE 11 – LOCATION ______________________________ (UTM or Six-Digit Grid Coordinate With MGRS Grid Zone Designator of Sighting)

LINE 12 – TIME ______________________________ (DTG of Sighting)

LINE 13 – STATUS ______________________________ (Status of Activity/Target)

LINE 14 – COUNT ______________________________ (Count of OB Items)

LINE 15 – TYPE ______________________________ (Type of OB Item)

LINE 16 – FURTHER ______________________________ (Follow-Up Report: Yes or No)

LINE 17 – WEATHER ______________________________ (Weather Conditions Over Target)

**Continued on next page.

TITLE: RECONNAISSANCE EXPLOITATION REPORT [RECCEXREP] (CONTINUED)
REPORT NUMBER: R020 {USMTF # C101}

LINE 18 – IMAGERY CONFIRMS ____________	(Imagery Confirmation of the Pilot Report/Debrief)
LINE 19 – SENSOR ____________	(Type of Sensor Used)
LINE 20 – COVERAGE ____________	(Percent of Target Covered. If less than 100 percent, explain why)
LINE 21 – NARRATIVE ____________	(Free Text for Additional Information Required for Clarification of Report)
LINE 22 – AUTHENTICATION ____________	(Report Authentication)

TITLE: RECONNAISSANCE NICKNAME REPORT [RECON 1]
REPORT NUMBER: R025 {USMTF # E631}

GENERAL INSTRUCTIONS: Use to report additions, changes, or deletions of operation order numbers and associated nicknames for reconnaissance. Reference: FM 34-2-1 and FM 101-5.

LINE 1 – DATE AND TIME ________________________________ (DTG)

LINE 2 – UNIT ________________________________ (Unit Making Report)

LINE 3 – TRANSACTION________________________________ (ADD, CHANGE, or DELETE)

LINE 4 – OPSORD________________________________ (Operations Order Identification Code)

LINE 5 – NEW NICKNAME________________________________ (Name of Newly Created or Replacement Nickname)

LINE 6 – OLD NICKNAME ________________________________ (Nickname Being Replaced or Deleted)

LINE 7 – EFFECTIVE ________________________________ (DTG Transaction Becomes Effective)

LINE 8 – AIRCRAFT TYPE ________________________________ (Type of Aircraft)

LINE 9 – PURPOSE________________________________ (IMINT, SIGINT, COMINT, or MASINT)

**Repeat lines 3 through 9 to identify multiple reconnaissance transactions. Assign sequential line numbers to succeeding iterations; for example, first iteration 3 through 9; second iteration 3a through 9a; third iteration 3b through 9b; and so on.

LINE 10 – NARRATIVE ________________________________ (Free Text for Additional Information Required for Clarification of Report)

LINE 11 – AUTHENTICATION ________________________________ (Report Authentication)

TITLE: RECONNAISSANCE TRACK REPORT [RECON 2]
REPORT NUMBER: R030 {USMTF # E632}

GENERAL INSTRUCTIONS: Use (1) to submit proposed reconnaissance tracks for mission approval and (2) to modify or delete existing active reconnaissance tracks. Reference: FM 34-2-1 and FM 101-5.

LINE 1 – DATE AND TIME ______________	(DTG)
LINE 2 – UNIT ______________	(Unit Making Report)
LINE 3 – TRANSACTION ______________	(REQUEST, CHANGE, or DELETE)
LINE 4 – NICKNAME ______________	(Mission Nickname)
LINE 5 – TRACK ______________	(Track Designator That Uniquely Identifies the Reconnaissance Track)
LINE 6 – DATE ______________	(DTG That the Reconnaissance Track is Submitted for JS Approval)
LINE 7 – OP AREA ______________	(Reconnaissance Operating Area)
LINE 8 – ASSESSMENT ______________	(Mission Assessment Code)
LINE 9 – GROUP ______________	(Track Approval Authority)
LINE 10 – TRACK EVENTS FOLLOW ______________	(Terrain Feature or Operational Graphic the Track Follows)
LINE 11 – SEQUENCE ______________	(Sequence Point or Event)
LINE 12 – EVENT ______________	(Event Code)
LINE 13 – LOCATION ______________	(UTM or Six-Digit Grid Coordinate With MGRS Grid Zone Designator)
LINE 14 – ALTITUDE ______________	(High- and Low-Flight Level Altitude Range)
LINE 15 – CLOSEST POINT ______________	(Closest Point of Approach in Whole Nautical Miles if Event Code is NE)

** Continued on next page.

TITLE: RECONNAISSANCE TRACK REPORT [RECON 2] (CONTINUED)
REPORT NUMBER: R030 {USMTF # E632}

LINE 16 – COUNTRY ______________________ (Country Code if Event Code is NE, TO, or LN.XX, if Unknown)

LINE 17 – CONTROL POINT ______________________ (No-Wind Time to Mission Control Point if Event Code is CP)

**Repeat lines 10 through 17 to report multiple reconnaissance track orbits. Assign sequential line numbers to succeeding iterations; for example, first iteration 10 through 17; second iteration 10a through 17a; third iteration 10b through 17b; and so on.

LINE 18 – ORBIT INFO FOLLOWS______________________ (Alerts receiver that the next four lines contain Orbit Information. When sending more than one track, state the number of the track being sent.)

EXAMPLE: ORBIT INFO FOLLOWS, TRACK ONE.

LINE 19 – ORDER______________________ (Order Number of Each Set of Events Starting With 1 at the First Enter Orbit Point)

LINE 20 – FROM ______________________ (First Sequence of Events Code at the Enter Orbit Point)

LINE 21 – TO ______________________ (Last Sequence of Events Code at the Exit Orbit Point)

LINE 22 – ORBITS______________________ (Number of Orbits That an Aircraft Makes; 1 Indicates No Orbits)

**Repeat lines 18 through 22 to report multiple reconnaissance track orbits. Assign sequential line numbers to succeeding iterations; for example, first iteration 18 through 22; second iteration 18a through 22a; third iteration 18b through 22b; and so on.

LINE 23 – AREA TIME______________________ (Total Time in Hours and Tenths That a Vehicle Spends Collecting Against Objectives)

LINE 24 – NARRATIVE ______________________ (Free Text for Additional Information Required for Clarification of Report)

LINE 25 – AUTHENTICATION ______________________ (Report Authentication)

TITLE: RECONNAISSANCE SCHEDULING REPORT [RECON 3]
REPORT NUMBER: R035 {USMTF # E633}

GENERAL INSTRUCTIONS: Use (1) to provide selected and specific data on proposed reconnaissance mission schedules, (2) to make changes to approved schedules, and (3) to inform of CINC/unit intention to fly specific reconnaissance missions. Reference: FM 34-2-1 and FM 101-5.

LINE 1 – DATE AND TIME	________________	(DTG)
LINE 2 – UNIT	________________	(Unit Making Report)
LINE 3 – TRANSACTION	________________	(SCHEDULE, ADD, CANCEL, CHANGE, DELAY, INDEFINITE DELAY, or RESCHEDULE)
LINE 4 – PERIOD	________________	(Mission Schedule Month and Year)
LINE 5 – REASON	________________	(Reason for Mission)
LINE 6 – MISSION NAME	________________	(Mission Nickname)
LINE 7 - MISSION DATA FOLLOWS	________________	(Alerts receiver that the next eleven lines contain Mission Data. When sending data for more than one mission, state the number of the mission being sent.)

EXAMPLE: MISSION DATA FOLLOWS; FIRST MISSION

LINE 8 – MISSION	________________	(Mission Number)
LINE 9 – TRACK	________________	(Mission Track Number)
LINE 10 – NUMBER TYPE	________________	(Count and Type of Aircraft)
LINE 11 – PURPOSE	________________	(Primary and Secondary Codes)
LINE 12 – TAKEOFF	________________	(Estimated Departure DTG Zulu)
LINE 13 – BASE	________________	(Departure Base ICAO Code)
LINE 14 – LANDING	________________	(Estimated Arrival DTG Zulu)
LINE 15 – BASE	________________	(Recovery Base ICAO Code)

**Continued on next page.

TITLE: RECONNAISSANCE SCHEDULING REPORT [RECON 3] (CONTINUED)
REPORT NUMBER: R035 {USMTF # E633}

LINE 16 – ASSESSMENT ______________________ (Mission Assessment Code)

LINE 17 – CONTINGENCY ______________________ (C if a Contingency Mission)

LINE 18 – COORDINATED ______________________ (C and 1 Through 9)

**Repeat lines 7 through 18 to report multiple missions. Assign sequential line numbers to succeeding iterations; for example, first iteration 7 through 18; second iteration 7a through 18a; third iteration 7b through 18b; and so on.

LINE 19 – TRACK OPTION DATA FOLLOWS ____________ (Alerts receiver that the next three lines contain Track Option Data. When sending data for more than one mission, state the number of the mission being sent.)

EXAMPLE: TRACK OPTION DATA FOLLOWS; FIRST TRACK

LINE 20 – MISSION ______________________ (Mission Number of Multitrack Mission Being Scheduled)

LINE 21 – OPTIONS ______________________ (Optional Track Numbers and Mission Assessment Codes)

**Repeat lines 19 through 21 to report multiple multitrack missions. Assign sequential line numbers to succeeding iterations; for example, first iteration 19 through 21; second iteration 19a through 21a; third iteration, 19b through 21b; and so on.

LINE 22 – COORDINATED MISSION DATA FOLLOWS__________ (Alerts receiver that the next three lines contain Coordinated Mission Data. If you are sending data for more than one mission, state the number of the mission being sent.)

EXAMPLE: COORDINATED MISSION DATA FOLLOWS; FIRST MISSION

LINE 23 – MISSION ______________________ (Mission Number of the Coordinated Mission Being Scheduled)

LINE 24 – WITH ______________________ (Up to 6 Coordinated Mission Numbers)

**Continued on next page.

TITLE: RECONNAISSANCE SCHEDULING REPORT [RECON 3] (CONTINUED)
REPORT NUMBER: R035 {USMTF # E633}

**Repeat lines 22 through 24 to report multiple multitrack missions. Assign sequential line numbers to succeeding iterations; for example, first iteration 22 through 24; second iteration 22a through 24a; third iteration 22b through 24b; and so on.

LINE 25 – SPECIAL DATA________________________________ (Enter data that does not belong in another line.)

LINE 26 – NARRATIVE ________________________________ (Free Text for Additional Information Required for Clarification of Report)

LINE 27 – AUTHENTICATION ____________________________ (Report Authentication)

TITLE: RECONNAISSANCE FOLLOWING REPORT [RECON 4]
REPORT NUMBER: R040 {USMTF # E634}

GENERAL INSTRUCTIONS: Use to provide specific information pertaining to a reconnaissance mission flight. Reference: FM 34-2-1, FM 100-103, and FM 101-5.

LINE 1 – DATE AND TIME	______________	(DTG)
LINE 2 – UNIT	______________	(Unit Making Report)
LINE 3 – TRANSACTION	______________	(LAUNCH, RECOVERY, RELAUNCH, or CHANGE)
LINE 4 – NICKNAME	______________	(Approved Nickname)
LINE 5 – MISSION	______________	(Mission Number)
LINE 6 – TRACK	______________	(Track Number)
LINE 7 – FORCE	______________	(Count and Type of Aircraft)
LINE 8 – ACTUAL DEPARTURE	______________	(Actual Departure DTG Zulu)
LINE 9 – BASE	______________	(Actual Launch Base ICAO Code or Coordinates)
LINE 10 – ESTIMATED ARRIVAL	______________	(Estimated Recovery DTG Zulu)
LINE 11 – BASE	______________	(Anticipated Recovery UTM or Six-Digit Grid Coordinate With MGRS Grid Zone Designator)
LINE 12 – ACTUAL ARRIVAL	______________	(Actual Landing DTG Zulu)
LINE 13 – BASE	______________	(Actual Recovery Base Using UTM or Six-Digit Grid Coordinate With MGRS Grid Zone Designator)
LINE 14 – STATUS	______________	(ABORT COMPLETE, ABORT INCOMPLETE, or AS SCHEDULED)
LINE 15 – REASON	______________	(Reason for Aborting Mission)

**Continued on next page.

TITLE: RECONNAISSANCE FOLLOWING REPORT [RECON 4] (CONTINUED)
REPORT NUMBER: R040 {USMTF # E634}

**Repeat lines 3 through 15 to report multiple multitrack missions. Assign sequential line numbers to succeeding iterations; for example, first iteration 3 through 15; second iteration 3a through 15a; third iteration 3b through 15b; and so on.

LINE 16 – NARRATIVE ____________________________________ (Free Text for Additional Information Required for Clarification of Report)

LINE 17 – AUTHENTICATION ______________________________ (Report Authentication)

TITLE: REQUEST CONFIRMATION [REQCONF]
REPORT NUMBER: R045 {USMTF # F657}

GENERAL INSTRUCTIONS: Use (1) to inform the requesting command, tasking authority, and tasked units of actions being taken and (2) to provide additional information about the missions. Use for both preplanned and immediate air-tasking cycles. Reference: FM 100-103 and FM 1-100.

LINE 1 – DATE AND TIME ______________________ (DTG)

LINE 2 – UNIT ______________________ (Unit Making Report)

LINE 3 – REQUEST ______________________ (Request Number)

LINE 4 – STATUS ______________________ (Approved or Disapproved)

**If disapproved, skip to LINE 15.

LINE 5 – MISSION ______________________ (Mission Number)

LINE 6 – PRIORITY ______________________ (Priority Assigned)

LINE 7 – TASKED ______________________ (Unit Providing Sorties)

LINE 8 – TYPE ______________________ (Mission Type)

LINE 9 – START ______________________ (DTG or Relative Time Aircraft Due to Arrive at Mission Location or Start Alert)

LINE 10 – STOP ______________________ (DTG or Relative Time Aircraft Due to Depart at Mission Location or Stop Alert)

LINE 11 – CALL SIGN ______________________ (Call Sign of the Aircraft Provided)

LINE 12 – NO. TYPE ______________________ (Number and Type Aircraft Provided)

LINE 13 – WEAPON ______________________ (Weapon Type)

LINE 14 – NARRATIVE ______________________ (Free Text for Additional Information Required for Clarification of Report)

LINE 15 – AUTHENTICATION ______________________ (Report Authentication)

TITLE: REQUEST FOR INFORMATION [RFI]
REPORT NUMBER: R050 {USMTF # F014}

GENERAL INSTRUCTIONS: Use to request information from requesting unit's command post or other units. Reference: FM 101-5.

LINE 1 – DATE AND TIME ______	(DTG)
LINE 2 – UNIT ______	(Unit Making Request)
LINE 3 – REQUEST ______	(Desired Information (Specific Order or Request))
LINE 4 – PRIORITY ______	(Requestor's Priority: ONE, TWO, THREE, or FOUR)
LINE 5 – BY ______	(DTG Information Required)
LINE 6 – LTIOV ______	(DTG of Latest Time of Intelligence/Information Value)
LINE 7 – NARRATIVE ______	(Free Text for Additional Information Required for Clarification of Report)
LINE 8 – AUTHENTICATION ______	(Report Authentication)

TITLE: RESPONSE TO REQUEST FOR INFORMATION [RRI]
REPORT NUMBER: R055 {USMTF # F015}

GENERAL INSTRUCTIONS: Use to reply to requests for information. If the information is contained in a previous message, the RRI should reference that message in Line 6. Reference FM 34-3 and FM 101-5.

LINE 1 – DATE AND TIME ______________________	(DTG)
LINE 2 – UNIT ______________________	(Unit Making Report)
LINE 3 – REQUEST______________________	(Request Number of Originator)
LINE 4 – RESPONSE______________________	(Answer to Requested Information)
LINE 5 – DTG ______________________	(DTG of Originator Request)
LINE 6 – NARRATIVE ______________________	(Free Text for Additional Information Required for Clarification of Report)
LINE 7 – AUTHENTICATION ______________________	(Report Authentication)

TITLE: ROAD CLEARANCE REQUEST [ROADCLRREQ]
REPORT NUMBER: R060 {USMTF # D827}

GENERAL INSTRUCTIONS: Use (1) to request road clearance and movement authority for the movement of outsized or overweight vehicles or (2) for the movement of convoys over a controlled or restricted road network. Reference: FM 100-15, FM 71-100, and FM 701-58.

LINE 1 – DATE AND TIME ______________________	(DTG)
LINE 2 – UNIT ______________________	(Unit Making Report)
LINE 3 – REFERENCE ______________________	(Reference: Data to Identify Request Being Referenced)
LINE 4 – UNIT ______________________	(Identifier or Designator of Component Unit or Agency Requesting Clearance on Making Move)
LINE 5 – PRIORITY ______________________	(Mission Priority Requested for Movement)
LINE 6 – CONVOY NAME ______________________	(Designated Convoy Name, if Any)
LINE 7 – COUNT OF SERIALS ______________________	(Count of Serials in Convoy)
LINE 8 – SERIAL TIME ______________________	(Convoy Serial Time Interval, In Minutes)
LINE 9 – COUNT OF MARCH GROUPS ______________________	(Count of March Groups in Convoy)
LINE 10 – MARCH GROUP TIME ______________________	(Convoy March Group Interval, in Minutes)
LINE 11 – HAZARDOUS CARGO ______________________	(Hazardous Cargo: YES or NO)
LINE 12 – START TIME ______________________	(Latest DTG and Time Zone Lead Vehicle is Planned to Pass Start Point)
LINE 13 – RATE OF MARCH ______________________	(Estimated Rate of March)
LINE 14 – START POINT ______________________	(UTM or Six-Digit Grid Coordinate With MGRS Grid Zone Designator of Starting Point)

**Continued on next page.

TITLE: ROAD CLEARANCE REQUEST [ROADCLRREQ] (CONTINUED)
REPORT NUMBER: R060 {USMTF # D827}

LINE 15 – START POC ________________	(Name of Start Point of Contact at Requesting Unit)
LINE 16 – LOCATION ________________	(Location of Start POC Using UTM or Six-Digit Grid Coordinate With MGRS Grid Zone Designator)
LINE 17 – FREQUENCIES ________________	(Start POC's Primary and Secondary Frequencies)
LINE 18 – PHONE NUMBERS ________________	(Start POC's Primary and Secondary Phone Numbers)
LINE 19 – ARRIVAL TIME ________________	(Estimated Arrival DTG of Moving Unit)
LINE 20 – END POINT ________________	(Location of Ending Point Using UTM Accurate to 100 Meters, Same as Line 16)
LINE 21 – END POC ________________	(Name of End Point of Contact at Requesting Unit)
LINE 22 – LOCATION ________________	(Location of End POC Using UTM or Six-Digit Grid Coordinate With MGRS Grid Zone Designator)
LINE 23 – FREQUENCIES ________________	(End POC's Primary and Secondary Frequencies)
LINE 24 – PHONE NUMBERS ________________	(End POC's Primary and Secondary Phone Numbers)
LINE 25 – ROUTE ________________	(Description of Route Desired by Requestor)
LINE 26 – VEHICLES ________________	(Number and Type of Vehicles in Movement)
LINE 27 – LENGTH ________________	(Length of All Outsized Vehicles, or Length of Longest Vehicle in Inches)

**Continued on next page.

TITLE: ROAD CLEARANCE REQUEST [ROADCLRREQ] (CONTINUED)
REPORT NUMBER: R060 {USMTF # D827}

LINE 28 – WIDTH ____________ (Width of All Outsized Vehicles or Width of Widest Vehicle in Inches)

LINE 29 – HEIGHT ____________ (Height of All Outsized Vehicles or Height of Highest Vehicle in Inches)

LINE 30 – WEIGHT ____________ (Weight of All Overweight Vehicles or Weight of Heaviest Vehicle)

LINE 31 – LOAD CLASS ____________ (Military Load Classification of Vehicles in Movement)

LINE 32 – HAZARDOUS CARGO DATA ____________ (Hazardous Cargo Classification, Shipping Name, Description, and Amount of Cargo)

LINE 33 – TRAFFIC CONTROL POINTS ____________ (Proposed Traffic Control Points for the Road Movement)

**Repeat lines 4 through 33 to request road clearance for more than one unit. Assign sequential line numbers to succeeding iterations; for example, first iteration 4 through 33; second iteration 4a through 33a; and third iteration 4b through 33b; and so on.

LINE 34 – NARRATIVE ____________ (Free Text for Additional Information Required for Clarification of Report)

LINE 35 – AUTHENTICATION ____________ (Report Authentication)

TITLE: ROUTE REPORT [ROUTEREP]
REPORT NUMBER: R065

GENERAL INSTRUCTIONS: Use to report results of route reconnaissance. Reference: FM 5-34, FM 17-95, and FM 17-98.

LINE 1 – DATE AND TIME ______________________ (DTG)

LINE 2 – UNIT ______________________ (Unit Making Report)

LINE 3 – RECON TIME ______________________ (DTG Start and Completion of Recon)

LINE 4 – RECOND ROUTE ______________________ (Route Reconned)

LINE 5 – TRAFFICABILITY ______________________ (Trafficability of Route)

LINE 6 – BUILT-UP AREAS ______________________ (Built-up Areas Along Route/Grid Coordinates)

LINE 7 – LATERAL ROUTES ______________________ (Lateral Routes Reconned/Results)

LINE 8 – BRIDGE CLASS ______________________ (Bridge Classification Report/Grid Coordinates)

LINE 9 – FORDS ______________________ (Fords and Crossings Site/Grid Coordinates)

LINE 10 – BYPASSES ______________________ (Overpasses, Underpasses, Culverts/Grid Coordinates)

LINE 11 – OBSTACLES ______________________ (UTM or Six-Digit Grid Coordinate With MGRS Grid Zone Designator of All Obstacles, Minefields, Contaminated Areas)

LINE 12 – ENEMY ______________________ (Enemy Activity That Can Influence Route/Grid Coordinates)

LINE 13 – NARRATIVE ______________________ (Free Text for Additional Information Required for Clarification of Report)

LINE 14 – AUTHENTICATION ______________________ (Report Authentication)

TITLE: RULES OF ENGAGEMENT AUTHORIZATION [ROEAUTH]
REPORT NUMBER: R070 {USMTF # F411}

GENERAL INSTRUCTIONS: Use to authorize or refuse the implementation of specific rules of engagement. Reference: FM 101-5.

LINE 1 – DATE AND TIME ______________________ (DTG)

LINE 2 – UNIT ______________________ (Unit Making Report)

LINE 3 – ROE APPROVED ______________________ (Approved Rules of Engagement and Any Applicable Constraints)

LINE 4 – ROE DENIED ______________________ (Rules of Engagement Being Denied)

LINE 5 – POLITICAL GUIDANCE ______________________ (Political Guidance Code for ROE Authorized)

LINE 6 – NARRATIVE ______________________ (Free Text for Additional Information Required for Clarification of Report)

LINE 7 – AUTHENTICATION ______________________ (Report Authentication)

TITLE: RULES OF ENGAGEMENT IMPLEMENTATION [ROEIMPL]
REPORT NUMBER: R075 {USMTF # F412}

GENERAL INSTRUCTIONS: Use to implement or cancel a specific rule of engagement. Reference: FM 101-5.

LINE 1 – DATE AND TIME	________________	(DTG)
LINE 2 – UNIT	________________	(Unit Making Report)
LINE 3 – ROE IMPLEMENT	________________	(Implemented Rules of Engagement With Constraints if Applicable)
LINE 4 – ROE CANCEL	________________	(Rules of Engagement Being Canceled)
LINE 5 – ROE SUMMARY	________________	(Summary of Rules of Engagement With Applicable Constraints in Effect Including Those Implemented by This Report)
LINE 6 – REPORTING POLICY	________________	(Reporting Policy Code)
LINE 7 – NARRATIVE	________________	(Free Text for Additional Information Required for Clarification of Report)
LINE 8 – AUTHENTICATION	________________	(Report Authentication)

TITLE: SAEDA REPORT [SAEDAREP]
REPORT NUMBER: S001

GENERAL INSTRUCTIONS: Use to provide notification of a Subversion and Espionage Directed Against the Army (SAEDA) incident. Reference: FM 34-60 and AR 310-50.

LINE 1 – DATE AND TIME ______________________________ (DTG)

LINE 2 – UNIT ______________________________ (Unit Making Report)

LINE 3 – INCIDENT TIME ______________________________ (DTG of Incident)

LINE 4 – LOCATION ______________________________ (UTM or Six-Digit Grid Coordinate With MGRS Grid Zone Designator)

LINE 5 – PERSONNEL ______________________________ (Personnel Involved)

a. WITNESSES ______________________________

b. SOURCES ______________________________

c. SUSPECTS ______________________________

LINE 6 – ACTIONS ______________________________ (Actions Taken by Personnel and Unit Involved)

LINE 7 – POINT OF CONTACT ______________________________ (Unit POC and Contact Information)

LINE 8 – NARRATIVE ______________________________ (Free Text for Additional Information Required for Clarification of Report)

LINE 9 – AUTHENTICATION ______________________________ (Report Authentication)

TITLE: SCATTERABLE MINEFIELD RECORD [SCATMINREC]
REPORT NUMBER: S005

GENERAL INSTRUCTIONS: Use to report emplacement of scatterable mines (SCATMINE). IAW unit SOPs or SCATMINE planning and execution policy, units will prepare and submit SCATMINREC in enough time to allow the request to be disseminated to all affected units. Once executed, it is critical to report each obstacle as a separate SCATMINREC to ensure it gains immediate visibility. (Placing SCATMINREC in the obstacle database, which is how most of the normal obstacles will be reported, slows dissemination.) This is especially important if the obstacle is on a unit boundary or beyond the FLOT. Reference: FM 20-32, FM 90-7, and FM 5-34.

LINE 1 – DATE AND TIME ________________________________ (DTG)

LINE 2 – UNIT ________________________________ (Unit Making Report)

LINE 3 – TGT/OBSTCL NO. ________________________________ (Target/Obstacle Number, as per Unit SOPs)

LINE 4 – EMPLACING SYSTEM ________________________________ (Emplacing System (Volcano (Air or Ground), Artillery (Type), MOPMS, Air Delivered (Air Force), or Gator)

LINE 5 – ANTITANK MINES ________________________________ (YES or NO)

LINE 6 – ANTIPERSONNEL MINES ________________________________ (YES or NO)

LINE 7 – LIFE CYCLE ________________________________ (DTG of Life Cycle Planned)

LINE 8 – AIM POINTS ________________________________ (Aim Points/Center Point of the Minefield)

a. ________________________________ (UTM or Six-Digit Grid of One Corner)

b. ________________________________ (UTM or Six-Digit Grid of One Corner)

c. ________________________________ (UTM or Six-Digit Grid of One Corner)

d. ________________________________ (UTM or Six-Digit Grid of One Corner)

**Continued on next page.

TITLE: SCATTERABLE MINEFIELD RECORD [SCATMINREC] (CONTINUED)
REPORT NUMBER: S005

LINE 9 – AIM POINTS ______________________ (Aim Points/Center Point of the Minefield)

a. ______________________ (UTM or Six-Digit Grid of One Corner)

b. ______________________ (UTM or Six-Digit Grid of One Corner)

c. ______________________ (UTM or Six-Digit Grid of One Corner)

d. ______________________ (UTM or Six-Digit Grid of One Corner)

LINE 10 – AIM POINTS ______________________ (Aim Points/Center Point of the Minefield)

a. ______________________ (UTM or Six-Digit Grid of One Corner)

b. ______________________ (UTM or Six-Digit Grid of One Corner)

c. ______________________ (UTM or Six-Digit Grid of One Corner)

d. ______________________ (UTM or Six-Digit Grid of One Corner)

LINE 11 – AIM POINTS ______________________ (Aim Points/Center Point of the Minefield)

a. ______________________ (UTM or Six-Digit Grid of One Corner)

b. ______________________ (UTM or Six-Digit Grid of One Corner)

c. ______________________ (UTM or Six-Digit Grid of One Corner)

d. ______________________ (UTM or Six-Digit Grid of One Corner)

**Continued on next page.

TITLE: SCATTERABLE MINEFIELD RECORD [SCATMINREC] (CONTINUED)
REPORT NUMBER: S005

LINE 12 – AIM POINTS ______________________ (Aim Points/Center Point of the Minefield)

LINE 13 – AIM POINTS ______________________ (Aim Points/Center Point of the Minefield)

LINE 14 – EMPLACING ______________________ (Unit Emplacing Mines and Report Number)

LINE 15 – SAFETY ZONE ______________________ (Size of Safety Zone)

LINE 16 – MINEFIELD MARKING ______________________ (Type of Marking, if Applicable)

LINE 17 – APPROVING AUTHORITY ______________________ (Approving Authority Commander)

LINE 18 – POC THIS REPORT ______________________ (Person Completing This Report)

LINE 19 – ACTIONS ______________________ (Actions Taken by Personnel Involved)

LINE 20 – NARRATIVE ______________________ (Free Text for Additional Information Required for Clarification of Report)

LINE 21 – AUTHENTICATION ______________________ (Report Authentication)

TITLE: SCATTERABLE MINEFIELD REQUEST [SCATMINREQ]
REPORT NUMBER: S010

GENERAL INSTRUCTIONS: Use to request authority to plan emplacement of scatterable mines (SCATMINE). IAW unit SOPs or the SCATMINE planning and execution policy, units will prepare and submit SCATMINREQ in enough time to allow the request to be staffed at the appropriate level and approval or disapproval returned to the requesting unit. Once a unit receives permission to plan a SCATMINE obstacle, it must still receive release authority before proceeding. This process is normally given when a scatterable minefield warning (SCATMINWARN) is sent 30 minutes prior to execution and the higher commander acknowledges and approves the release. Once the minefield is in place, a minefield/obstacle report (SCATMINREC) is sent to register the minefield. This is key, as the minefield may be on a unit boundary or beyond the FLOT. Reference: FM 20-32, FM 90-7, and FM 5-34.

LINE 1 – DATE AND TIME ______________________ (DTG)

LINE 2 – UNIT ______________________ (Unit Making Report)

LINE 3 – TGT/OBSTCL NO. ______________________ (Target/Obstacle Number, as per Unit SOPs)

LINE 4 – EMPLACING SYSTEM ______________________ (Emplacing System (Volcano (Air or Ground), Artillery (Type), MOPMS, Air Delivered (Air Force), or Gator)

LINE 5 – ANTITANK MINES ______________________ (YES or NO)

LINE 6 – ANTIPERSONNEL MINES ______________________ (YES or NO)

LINE 7 – ATTITUDE ______________________ (Attitude of Minefield)

LINE 8 – DIMENSIONS ______________________ (Dimensions of Minefield)

LINE 9 – AIM POINTS ______________________ (Number of Aim Points/Corner Points With Grid Coordinates)

LINE 10 – SAFETY ZONE ______________________ (Size of Safety Zone)

LINE 11 – MINEFIELD MARKING ______________________ (Type of Marking, if Applicable)

LINE 12 – UNIT OBSERVING ______________________ (Unit Observing)

LINE 13 – MISSION ______________________ (Task, Purpose and Intent, if Required)

LINE 14 – LIFE CYCLE ______________________ (DTG of Life Cycle Planned)

**Continued on next page.

TITLE: SCATTERABLE MINEFIELD REQUEST [SCATMINREQ] (CONTINUED)
REPORT NUMBER: S010

LINE 15 – ACTIONS ________________________________ (Actions Taken by Personnel Involved)

LINE 16 – NARRATIVE ______________________________ (Free Text for Additional Information Required for Clarification of Report)

LINE 17 – AUTHENTICATION _________________________ (Report Authentication)

TITLE: SCATTERABLE MINEFIELD WARNING (SCATMINWARN)
REPORT NUMBER: S015

GENERAL INSTRUCTIONS: Use to request authority to execute a planned scatterable minefield (SCATMINE) obstacle. Use the SCATMINREQ to request authority to plan a SCATMINE obstacle and the SCATMINREC to record an executed SCATMINE obstacle. Reference: FM 20-32, FM 90-7, and FM 5-34.

LINE 1 – DATE AND TIME ______________	(DTG)
LINE 2 – UNIT ______________	(Unit Making Report)
LINE 3 – TGT/OBSTCL NO. ______________	(Target/Obstacle Number, as per Unit SOPs)
LINE 4 – EMPLACING SYSTEM ______________	(Emplacing System (Volcano (Air or Ground), Artillery (Type), MOPMS, Air Delivered (Air Force), or Gator)
LINE 5 – ANTITANK MINES ______________	(Antitank Mines (YES or NO))
LINE 6 – ANTIPERSONNEL MINES ______________	(Antipersonnel Mines (YES or NO))
LINE 7 – AIMING POINTS ______________	(Grid Coordinates of Aim Points/Corner Points, if Required Due to Refinement When Authorized)
LINE 8 – SAFETY ZONE ______________	(Size of Safety Zone)
LINE 9 – MINEFIELD MARKING ______________	(Type of Marking, if Applicable)
LINE 10 – LIFE CYCLE ______________	(DTG of Life Cycle)
LINE 11 – ACTIONS ______________	(Actions Taken by Personnel Involved)
LINE 12 – NARRATIVE ______________	(Free Text for Additional Information Required for Clarification of Report)
LINE 13 – AUTHENTICATION ______________	(Report Authentication)

TITLE: SEARCH AND RESCUE INCIDENT REPORT [SARIR]
REPORT NUMBER: S020 {USMTF # C482}

GENERAL INSTRUCTIONS: Use to report any situation that may require a search and rescue effort. Reference: FM 1-108 and FM 1-513.

LINE 1 – DATE AND TIME ______________________	(DTG)
LINE 2 – UNIT ______________________	(Unit Making Report)
LINE 3 – CALL SIGN ______________________	(Call Sign of Disabled or Lost Aircraft, Ship, Submarine, or Other)
LINE 4 – TYPE ______________________	(Type of Disabled or Lost Aircraft, Ship, Submarine, or Other)
LINE 5 – COLOR ______________________	(Color of Disabled or Lost Aircraft, Ship, Submarine, or Other)
LINE 6 – ID ______________________	(Aircraft Tail or Side Number, Ship/Submarine Hull Number, or Other Number)
LINE 7 – LOCATION ______________________	(Location Disabled or Lost Aircraft, Ship, Submarine, or Other in UTM or Six-Digit Grid Coordinate With MGRS Grid Zone Designator)
LINE 8 – QUALIFIER ______________________	(Location Qualifier: ACTUAL or ESTIMATED Followed by LAND or SEA)
LINE 9 – TIME OF INCIDENT ______________________	(DTG of Incident)
LINE 10 – CAUSE ______________________	(Cause of Disabled or Lost Aircraft, Ship, Submarine, or Other)
LINE 11 – PERSONNEL ______________________	(Count of Personnel on Board and Qualifier: ACTUAL or ESTIMATED)
LINE 12 – STATUS ______________________	(Enter Count of Personnel and Their Statuses)

**Continued on next page.

TITLE: SEARCH AND RESCUE INCIDENT REPORT [SARIR] (CONTINUED)
REPORT NUMBER: S020 {USMTF # C482}

LINE 13 – REQUIRE ______________________________ (Enter RCC or COMBINED ASSISTANCE if SAR Assistance is Required)

LINE 14 – POINT OF CONTACT ______________________ (Enter the Point of Contact and Telephone Number)

LINE 15 – NARRATIVE ____________________________ (Free Text for Additional Information Required for Clarification of Report)

LINE 16 – AUTHENTICATION _______________________ (Report Authentication)

TITLE: SEARCH AND RESCUE (SAR) REQUEST [SARREQ]
REPORT NUMBER: S025 {USMTF # D669}

GENERAL INSTRUCTIONS: Use to request forces to participate in a search and rescue (SAR) mission. Reference: FM 1-108 and FM 1-513.

LINE 1 – DATE AND TIME ______________________ (DTG)

LINE 2 – UNIT ______________________ (Unit Making Report)

LINE 3 – REQUEST______________________ (Request Number)

LINE 4 – PRIORITY ______________________ (Priority of Request: 1, 1A through 1Z; 2, 2A through 2Z; 3, 3A Through 3Z; 4, 4A through 4Z)

LINE 5 – TYPE______________________ (Air Mission Type)

LINE 6 – ON TIME ______________________ (DTG Aircraft Are Requested to be at SAR Location or ASAP)

LINE 7 – OFF TIME______________________ (DTG Aircraft Are to Depart From SAR Location)

LINE 8 – NO. TYPE______________________ (Number and Type of Assets Requested)

LINE 9 – LOCATION ______________________ (SAR Location in UTM or Six-Digit Grid Coordinate With MGRS Grid Zone Designator)

LINE 10 – NARRATIVE ______________________ (Free Text for Additional Information Required for Clarification of Report)

LINE 11 – AUTHENTICATION ______________________ (Report Authentication)

TITLE: SEARCH AND RESCUE SITUATION SUMMARY REPORT [SARSIT]
REPORT NUMBER: S030 {USMTF # C420}

GENERAL INSTRUCTIONS: Use to coordinate, summarize, or terminate joint search and rescue operations. Reference: FM 1-108 and FM 1-513.

LINE 1 – DATE AND TIME ______________________	(DTG)
LINE 2 – UNIT ______________________	(Unit Making Report)
LINE 3 – MISSION NUMBER ______________________	(Enter the JRCC SAR Mission Number)
LINE 4 – STATUS ______________________	(SAR Status: COMPLETED or TERMINATED if SAR Activity Has Ceased and Will Not be Resumed at a Later Time; SUSPENDED, if SAR Activity is Discontinued and Objective is Not Recovered)
LINE 5 – CALL SIGN ______________________	(Call Sign of Disabled or Lost Aircraft, Ship, Submarine, or Other)
LINE 6 – TYPE ______________________	(Type of Disabled or Lost Aircraft, Ship, Submarine, or Other)
LINE 7 – LOCATION ______________________	(UTM or Six-Digit Grid Coordinate With MGRS Grid Zone Designator of SAR Incident)
LINE 8 – PERSONNEL ______________________	(Number of Personnel Involved in Incident)
LINE 9 – PERSONNEL STATUS ______________________	(Status of Personnel Involved in Incident (RECOVERED))
LINE 10 – NARRATIVE ______________________	(Free Text for Additional Information Required for Clarification of Report)
LINE 11 – AUTHENTICATION ______________________	(Report Authentication)

TITLE: SENSITIVE ITEMS REPORT [SENITREP]
REPORT NUMBER: S035

GENERAL INSTRUCTIONS: Use to convey status of military equipment deemed sensitive by chain of command. Send IAW unit SOPs. Reference: FM 19-30.

LINE 1 – DATE AND TIME	______________	(DTG)
LINE 2 – UNIT	______________	(Unit Making Report)
LINE 3 – STATUS	______________	(UP-100 percent Complete or UNIT GREEN to Denote Report With No Discrepancy; if Unit is Green Skip to Line 10; When Discrepancy is Noted Report UNIT RED and Continue Report)
LINE 4 – LOSING UNIT	______________	(Unit Making Report Loss)
LINE 5 – ITEM	______________	(Item Lost by Serial Number)
LINE 6 – DTG OF LOSS	______________	(DTG of Loss)
LINE 7 – DETAILS	______________	(Circumstances of Loss)
LINE 8 – ACTION TAKEN	______________	(Actions Taken to Recover the Item)
LINE 9 – NARRATIVE	______________	(Free Text for Additional Information Required for Clarification of Report)
LINE 10 – AUTHENTICATION	______________	(Report Authentication)

TITLE: SEVERE WEATHER WARNING [SVRWXWARN]
REPORT NUMBER: S040

GENERAL INSTRUCTIONS: Use to warn commanders of severe weather affecting the area of operations. Transmit this message as required until normal weather communications are established. Reference: FM 34-81.

LINE 1 – DATE AND TIME	__________	(DTG)
LINE 2 – UNIT	__________	(Unit Making Report)
LINE 3 – VALID	__________	(DTG the Warning is Effective From)
LINE 4 – UNTIL	__________	(DTG the Warning is Effective to)
LINE 5 – WEATHER	__________	(Atmosphere Conditions With Respect to Cloudiness, Precipitation, or Other Weather Phenomena)
LINE 6 – VISIBILITY	__________	(Minimum Visibility in Meters)
LINE 7 – CEILING	__________	(Lowest Forecast Ceiling in Feet)
LINE 8 – SPEED	__________	(Maximum Sustained Wind Speed in Knots)
LINE 9 – GUSTS	__________	(Peak Wind Gusts in Knots)
LINE 10 – DIRECTION	__________	(Wind Direction in Degrees)
LINE 11 – NARRATIVE	__________	(Free Text for Additional Information Required for Clarification of Report)
LINE 12 – AUTHENTICATION	__________	(Report Authentication)

TITLE: SLANT REPORT [SLANTREP]
REPORT NUMBER: S045

GENERAL INSTRUCTIONS: Use to give the commander accurate and routine information regarding status of critical personnel and equipment necessary for the unit's operation success. Submit when necessary or as directed. The commander designates the information to report during planning or by unit SOPs. Reference: FM 71-2 and FM 71-3.

LINE 1 – DATE AND TIME ________________________________ (DTG)

LINE 2 – UNIT ________________________________ (Unit Making Report)

LINE 3 – CDR'S INFO REQ ________________________________ (Commander's Information Requirement)

LINE 4 – EQUIPMENT ________________________________

LINE 5 – PERSONNEL ________________________________

**Unit SOPs determine the items and the order they are reported in Lines 4 and 5.

EXAMPLE: M1A2/M2A3/AH-64/MLRS/19K. "Sent As 24 of 38/38 of 38/16 of 24/389 of 400"

LINE 6 – NARRATIVE ________________________________ (Free Text for Additional Information Required for Clarification of Report)

LINE 7 – AUTHENTICATION ________________________________ (Report Authentication)

TITLE: SORTIE ALLOTMENT [SORTIEALOT]
REPORT NUMBER: S050 {USMTF # A656}

GENERAL INSTRUCTIONS: Use to provide a means by which a joint force commander approves the air employment/allocation plans of his subordinate commanders and fills his subordinate commanders' requests from sorties declared in excess in the subordinate command's Air Allocation/Request Message. Reference: JP 3-56.1.

LINE 1 – DATE AND TIME ________________ (DTG)

LINE 2 – UNIT ________________ (Unit Making Report)

LINE 3 – EMPLOYMENT PLAN ________________ (Significant Changes to Applicable Component's Employment Plan)

LINE 4 – SUPPORT RECEIVED________________ (Addressees Requests Filled; All Others Disapproved)

LINE 5 – REQUEST FILLED BY ________________ (Request Number and Component Filling Request)

LINE 6 – REQUEST FILLED BY ________________ (Request Number and Component Filling Request)

LINE 7 – REQUEST FILLED BY ________________ (Request Number and Component Filling Request)

LINE 8 – SUPPORT GIVEN ________________ (Requests to be Filled by Addressee)

LINE 9 – REQUEST________________ (Request Number)

LINE 10 – DATA ________________ (Mission Type, Start and Stop Times, Number of Sorties Requested)

LINE 11 – REQUEST________________ (Request Number)

LINE 12 – DATA ________________ (Mission Type, Start and Stop Times, Number of Sorties Requested)

LINE 13 – NARRATIVE ________________ (Free Text for Additional Information Required for Clarification of Report)

LINE 14 – AUTHENTICATION ________________ (Report Authentication)

TITLE: SPILL REPORT [SPILLREP]
REPORT NUMBER: S055

GENERAL INSTRUCTIONS: Use to send timely information on status of oil, hazardous material, or hazardous waste spill that could have immediate environmental and/or health effects. Sent IAW unit SOPs and commander's direction. NOTE: Spill reporting and reportable quantities are mandated by federal law.

LINE 1 – DATE AND TIME	______	(DTG)
LINE 2 – UNIT	______	(Unit Making Report)
LINE 3 – DATE/TIME	______	(DTG of Spill Discovery)
LINE 4 – MATERIAL	______	(Material Spilled)
LINE 5 – QUANTITY	______	(Quantity of Spilled Material)
LINE 6 – LOCATION	______	(UTM or Six-Digit Grid Coordinate With MGRS Grid Zone Designator of Spill)
LINE 7 – CAUSE	______	(Cause and Supervising Unit)
LINE 8 – SIZE	______	(Size of Affected Area)
LINE 9 – DAMAGE	______	(Damage to the Natural Environment (if required))
LINE 10 – HAZARDS	______	(Hazards to Friendly Forces and/or Civilian Personnel)
LINE 11- ACTIONS	______	(Summary of Actions Taken)
LINE 12 – UNIT POC	______	(Supervising Unit POC)
LINE 13 – ASSISTANCE	______	(Assistance Required/Requested)
LINE 14 – NARRATIVE	______	(Free Text for Additional Information Required for Clarification of Report)
LINE 15 – AUTHENTICATION	______	(Report Authentication)

TITLE: SPOT REPORT [SPOTREP]
REPORT NUMBER: S060

GENERAL INSTRUCTIONS: Use to send information to provide timely intelligence or status regarding events that could have an immediate and significant effect on current planning and operations.
Reference: FM 17-15, FM 17-98, and FM 7-7.

LINE 1 – DATE AND TIME ____________________ (DTG)

LINE 2 – UNIT ____________________ (Unit Making Report)

LINE 3 – SIZE ____________________ (Size of Enemy Unit)

LINE 4 – ACTIVITY ____________________ (Enemy Activity at DTG of Report)

LINE 5 – LOCATION ____________________ (UTM or Six-Digit Grid Coordinate With MGRS Grid Zone Designator of Enemy Activity or Event Observed)

LINE 6 – UNIT ____________________ (Enemy Unit)

LINE 7 – TIME ____________________ (DTG of Observation)

LINE 8 – EQUIPMENT ____________________ (Equipment of Unit Observed)

LINE 9 – SENDER'S ASSESSMENT ____________________ (Specific Sender Information)

LINE 10 – NARRATIVE ____________________ (Free Text for Additional Information Required for Clarification of Report)

LINE 11 – AUTHENTICATION ____________________ (Report Authentication)

TITLE: STOP JAMMING MESSAGE [STOPJAMMING]
REPORT NUMBER: S065 {USMTF # D675}

GENERAL INSTRUCTIONS: Use to terminate a jamming task being conducted by an EA asset. Reference: FM 34-40.

LINE 1 – DATE AND TIME ______________________ (DTG)

LINE 2 – UNIT ______________________ (Unit Making Report)

LINE 3 – FREQUENCY ______________________ (Enter the Radio Frequency or Enter ALL if Jamming is to Stop on All Frequencies)

LINE 4 – NARRATIVE ______________________ (Free Text for Additional Information Required for Clarification of Report)

LINE 5 – AUTHENTICATION ______________________ (Report Authentication)

TITLE: STRAGGLER STATUS REPORT [STRAGSTATREP]
REPORT NUMBER: S070

GENERAL INSTRUCTIONS: Use to report any personnel who are in areas not authorized by current standing orders and are not assessed as missing in action. Reference: FM 19-1.

LINE 1 – DATE AND TIME ______________________ (DTG)

LINE 2 – UNIT ______________________ (Unit Making Report)

LINE 3 – NUMBER ______________________ (Number of Stragglers by Rank and Unit)

LINE 4 – LOCATION ______________________ (Location of Stragglers Picked Up: UTM or Six-Digit Grid Coordinate With MGRS Grid Zone Designator)

LINE 5 – AA ______________________ (Disposition/Location of Straggler AA)

LINE 6 – NARRATIVE ______________________ (Free Text for Additional Information Required for Clarification of Report)

LINE 7 – AUTHENTICATION ______________________ (Report Authentication)

TITLE: SUMMARY REPORT OF NUCLEAR DETONATIONS [NUDETSUM]
REPORT NUMBER: S075 {USMTF # C442}

GENERAL INSTRUCTIONS: Use to report information pertaining to nuclear detonations on areas outside the North American continent. This report will support damage assessment at the national level. Reference: FM 3-14.

LINE 1 – DATE AND TIME ________________________ (DTG)

LINE 2 – UNIT ________________________ (Unit Making Report)

LINE 3 – AS OF ________________________ (As of DTG for Applicability of NUDETSUM Information)

LINE 4 – SEQUENCE ________________________ (Serial Number Assigned to This Nuclear Detonation)

LINE 5 – PLACE ________________________ (Place or Area Name of the Nuclear Detonation)

LINE 6 – COORDINATES________________________ (Detonation's UTM or Six-Digit Grid Coordinate With Grid Zone Designator)

LINE 7 – DETONATION________________________ (DTG Detonation Occurred)

**Repeat lines 4 through 7 to report all observed nuclear detonations.

LINE 8 – NARRATIVE ________________________ (Free Text for Additional Information Required for Clarification of Report)

LINE 9 – AUTHENTICATION ________________________ (Report Authentication)

TITLE: SUPPORT-AIR CORRIDOR [SPRT.AIRCOR]
REPORT NUMBER: S080 {USMTF # A263}

GENERAL INSTRUCTIONS: Use to report, modify, or cancel an airspace coordination area to fire support agencies. Reference: FM 100-103.

LINE 1 – DATE AND TIME ________________ (DTG)

LINE 2 – UNIT ________________ (Unit Making Report)

LINE 3 – POI ________________ (Primary Option Indicator. Enter ADD, AMEND, or CANCEL)

LINE 4 – PLAN ________________ (Fire Plan Designation)

LINE 5 – NAME ________________ (Air Corridor Name)

LINE 6 – FROM ________________ (DTG Effective From)

LINE 7 – TO ________________ (DTG Effective To)

LINE 8 – ZONE ALFA ________________ (UTM Grid Zone and 100-KM Square)

LINE 9 – POINT ALFA ________________ (UTM 1-Meter Easting and UTM 1-Meter Northing)

LINE 10 – ALFA WIDTH ________________ (Width in Meters)

LINE 11 – ALFA MIN ________________ (Minimum Altitude in Meters)

LINE 12 – ALFA MAX ________________ (Maximum Altitude in Meters)

LINE 13 – ZONE BRAVO ________________ (UTM Grid Zone and 100-KM Square)

LINE 14 – POINT BRAVO ________________ (UTM 1-Meter Easting and UTM 1-Meter Northing)

LINE 15 – BRAVO WIDTH ________________ (Width in Meters)

LINE 16 – BRAVO MIN ________________ (Minimum Altitude in Meters)

LINE 17 – BRAVO MAX ________________ (Maximum Altitude in Meters)

**Continued on next page.

TITLE: SUPPORT-AIR CORRIDOR [SPRT.AIRCOR] (CONTINUED)
REPORT NUMBER: S080 {USMTF # A263}

LINE 18 – ZONE CHARLIE ________________ (UTM Grid Zone and 100-KM Square)

LINE 19 – POINT CHARLIE ________________ (UTM 1-Meter Easting and UTM 1-Meter Northing)

LINE 20 – CHARLIE WIDTH ________________ (Width in Meters)

LINE 21 – CHARLIE MIN ________________ (Minimum Altitude in Meters)

LINE 22 – CHARLIE MAX ________________ (Maximum Altitude in Meters)

LINE 23 – ZONE DELTA ________________ (UTM Grid Zone and 100-KM Square)

LINE 24 – POINT DELTA ________________ (UTM 1-Meter Easting and UTM 1-Meter Northing)

LINE 25 – DELTA WIDTH ________________ (Width in Meters)

LINE 26 – DELTA MIN ________________ (Minimum Altitude in Meters)

LINE 27 – DELTA MAX ________________ (Maximum Altitude in Meters)

LINE 28 – ZONE ECHO ________________ (UTM Grid Zone and 100-KM Square)

LINE 29 – POINT ECHO ________________ (UTM 1-Meter Easting and UTM 1-Meter Northing)

LINE 30 – ECHO WIDTH ________________ (Width in Meters)

LINE 31 – ECHO MIN ________________ (Minimum Altitude in Meters)

LINE 32 – ECHO MAX ________________ (Maximum Altitude in Meters)

LINE 33 – ZONE FOXTROT ________________ (UTM Grid Zone and 100-KM Square)

LINE 34 – POINT FOXTROT ________________ (UTM 1-Meter Easting and UTM 1-Meter Northing)

LINE 35 – FOXTROT WIDTH ________________ (Width in Meters)

**Continued on next page.

TITLE: SUPPORT-AIR CORRIDOR [SPRT.AIRCOR] (CONTINUED)
REPORT NUMBER: S080 {USMTF # A263}

LINE 36 – FOXTROT MIN ______________	(Minimum Altitude in Meters)
LINE 37 – FOXTROT MAX ______________	(Maximum Altitude in Meters)
LINE 38 – ZONE GOLF ______________	(UTM Grid Zone and 100-KM Square)
LINE 39 – POINT GOLF ______________	(UTM 1-Meter Easting and UTM 1-Meter Northing)
LINE 40 – GOLF WIDTH ______________	(Width in Meters)
LINE 41 – GOLF MIN ______________	(Minimum Altitude in Meters)
LINE 42 – GOLF MAX ______________	(Maximum Altitude in Meters)
LINE 43 – ZONE HOTEL ______________	(UTM Grid Zone and 100-KM Square)
LINE 44 – POINT HOTEL ______________	(UTM 1-Meter Easting and UTM 1-Meter Northing)
LINE 45 – NARRATIVE ______________	(Free Text for Additional Information Required for Clarification of Report)
LINE 46 – AUTHENTICATION ______________	(Report Authentication)

TITLE: SUPPORT-BATTLEFIELD GEOMETRY [SPRT.GEOM]
REPORT NUMBER: S085 {USMTF # A262}

GENERAL INSTRUCTIONS: Use to transfer, amend, cancel, or propose any battlefield geometry (which includes fire support coordination measures) in support of land combat operations. Reference: FM 100-103.

LINE 1 – DATE AND TIME	______	(DTG)
LINE 2 – UNIT	______	(Unit Making Report)
LINE 3 – POI	______	(Primary Option Indicator. Enter ADD, AMEND, or CANCEL)
LINE 4 – SOI	______	(Secondary Option Indicator. Enter ADDED POINTS if ADD is in Line 1 and Additional Points Will Follow in Another Report)
LINE 5 – PLAN	______	(Fire Plan Designation)
LINE 6 – FROM	______	(DTG Effective From)
LINE 7 – TO	______	(DTG Effective to)
LINE 8 – BGT	______	(Battlefield Geometry Type)
LINE 9 – NAME	______	(Battlefield Geometry Type Name)
LINE 10 – AUTHORITY	______	(Coordination/Establishing Authority)
LINE 11 – ONE GRID	______	(UTM or Six-Digit Grid Coordinate With MGRS Grid Zone Designator)
LINE 12 – TWO GRID	______	(UTM or Six-Digit Grid Coordinate With MGRS Grid Zone Designator)
LINE 13 – THREE GRID	______	(UTM or Six-Digit Grid Coordinate With MGRS Grid Zone Designator)

**Continued on next page.

TITLE: SUPPORT-BATTLEFIELD GEOMETRY [SPRT.GEOM] (CONTINUED)
REPORT NUMBER: S085 {USMTF # A262}

LINE 14 – FOUR GRID ________	(UTM or Six-Digit Grid Coordinate With MGRS Grid Zone Designator)
LINE 15 – FIVE GRID ________	(UTM or Six-Digit Grid Coordinate With MGRS Grid Zone Designator)
LINE 16 – CIRCLE ________	(UTM Grid Zone and 100-KM Square, UTM 1-Meter Easting, and UTM 1-Meter Northing)
LINE 17 – RADIUS ________	(Circular Area Radius in Meters)
LINE 18 – AMMO-RESTRICTED ________	(Restrictive Fire Area Restricted Ammo Indicator)
LINE 19 – AMMO-RESTRICTED ________	(Restrictive Fire Area Restricted Ammo Indicator)
LINE 20 – BOUNDARY POINT ________	(Coordinate Point Sequence Number)
LINE 21 – NAME ________	(Boundary Name)
LINE 22 – BOUNDARY POINT ________	(Coordinate Point Sequence Number)
LINE 23 – NAME ________	(Boundary Name)
LINE 24 – NARRATIVE ________	(Free Text for Additional Information Required for Clarification of Report)
LINE 25 – AUTHENTICATION ________	(Report Authentication)

TITLE: SUPPORT-DAMAGE AVOIDANCE AREA [SPRT.DAACAT]
REPORT NUMBER: S090 {USMTF # A280}

GENERAL INSTRUCTIONS: Use (1) to establish nuclear damage categories and (2) to disseminate command guidance establishing chemical avoidance criteria within damage avoidance areas. Reference: FM 3-3-1.

LINE 1 – DATE AND TIME ______________________________	(DTG)
LINE 2 – UNIT ______________________________	(Unit Making Report)
LINE 3 – POI ______________________________	(Primary Option Indicator. Enter ADD, AMEND, or CANCEL)
LINE 4 – PLAN ______________________________	(Fire Plan Designation)
LINE 5 – AREA NAME ______________________________	(Damage Avoidance Area Name)
LINE 6 – ZONE ______________________________	(UTM Grid Zone and 100-KM Square)
LINE 7 – ONE GRID ______________________________	(UTM 1-Meter Easting, UTM 1-Meter Northing)
LINE 8 – TWO GRID ______________________________	(UTM 1-Meter Easting, UTM 1-Meter Northing)
LINE 9 – THREE GRID ______________________________	(UTM 1-Meter Easting, UTM 1-Meter Northing)
LINE 10 – FOUR GRID ______________________________	(UTM 1-Meter Easting, UTM 1-Meter Northing)
LINE 11 – FIVE GRID ______________________________	(UTM 1-Meter Easting, UTM 1-Meter Northing)
LINE 12 – SIX GRID ______________________________	(UTM 1-Meter Easting, UTM 1-Meter Northing)
LINE 13 – SEVEN GRID ______________________________	(UTM 1-Meter Easting, UTM 1-Meter Northing)
LINE 14 – EIGHT GRID ______________________________	(UTM 1-Meter Easting, UTM 1-Meter Northing)

**Continued on next page.

TITLE: SUPPORT-DAMAGE AVOIDANCE AREA [SPRT.DAACAT] (CONTINUED)
REPORT NUMBER: S090 {USMTF # A280}

LINE 15 – NINE GRID	______________	(UTM 1-Meter Easting, UTM 1-Meter Northing)
LINE 16 – CIRCLE	______________	(UTM 1-Meter Easting, UTM 1-Meter Northing)
LINE 17 – RADIUS	______________	(Circular Area Radius in Meters)
LINE 18 – PERSONNEL	______________	(Personnel Damage Category and Level of Assurance)
LINE 19 – FACILITY	______________	(Facility Damage Category and Level of Assurance)
LINE 20 – THERMAL	______________	(Thermal Ignition Damage Category and Level of Assurance)
LINE 21 – CHEMICAL MINIMUM	______________	(Minimum Desired Chemical Effects)
LINE 22 – MAXIMUM	______________	(Maximum Desired Chemical Effects and Persistent Indicator)
LINE 23 – NARRATIVE	______________	(Free Text for Additional Information Required for Clarification of Report)
LINE 24 – AUTHENTICATION	______________	(Report Authentication)

TITLE: SURVEILLANCE AND RECONNAISSANCE PLAN REPORT [SURRECONREP]
REPORT NUMBER: S095

GENERAL INSTRUCTIONS: Use to report reconnaissance and surveillance plans. Reference: FM 34-2-1.

LINE 1 – DATE AND TIME ______________________________ (DTG)

LINE 2 – UNIT ______________________________ (Unit Making Report)

LINE 3 – OPERATION ______________________________ (Period and Operation Covered)

LINE 4 – PIR/IR ______________________________ (Identify the Collection Emphasis by Type and by the CCIR (PIR/IR) it is Assigned, Include the Task and Purpose of Collector to IR)

LINE 5 – EMPHASIS ______________________________ (Collection Emphasis by Discipline)

a. SIGINT ______________________________

b. HUMINT ______________________________

c. IMINT ______________________________

LINE 6 – SORs ______________________________

LINE 7 – NARRATIVE ______________________________ (Free Text for Additional Information Required for Clarification of Report)

LINE 8 – AUTHENTICATION ______________________________ (Report Authentication)

TITLE: SURVEY-CONTROL POINT ACCESS REQUEST [SURV.TPAC]
REPORT NUMBER: S100 {USMTF # D272}

GENERAL INSTRUCTIONS: Use to retrieve survey control points by circular, thrust line, or rectangular search. The report provides a means to request survey control point data. These control points provide potential starting and closing points for the extension of control. Reference: FM 6-2.

LINE 1 – DATE AND TIME	____________________	(DTG)
LINE 2 – UNIT	____________________	(Unit Making Report)
LINE 3 – ORDER	____________________	(Order of Survey Accuracy)
LINE 4 – ZONE	____________________	(UTM Grid Zone and 100-KM Square)
LINE 5 – CIRCLE	____________________	(UTM 1-Meter Easting and UTM 1-Meter Northing)
LINE 6 – RADIUS	____________________	(Circular Area Radius, Survey Search)
LINE 7 – ONE GRID	____________________	(UTM 1-Meter Easting, UTM 1-Meter Northing)
LINE 8 – TWO GRID	____________________	(UTM 1-Meter Easting, UTM 1-Meter Northing)
LINE 9 – WIDTH	____________________	(Width of Thrust Line in Meters)
LINE 10 – NARRATIVE	____________________	(Free Text for Additional Information Required for Clarification of Report)
LINE 11 – AUTHENTICATION	____________________	(Report Authentication)

TITLE: SURVEY-CONTROL POINT STORAGE (INPUT/OUTPUT) MESSAGE [SURV.SCPST]
REPORT NUMBER: S105 {USMTF # C270}

GENERAL INSTRUCTIONS: Use to establish, cancel, or transmit SCP data. Reference: FM 6-2.

LINE 1 – DATE AND TIME ________________	(DTG)
LINE 2 – UNIT ________________	(Unit Making Report)
LINE 3 – POI ________________	(Primary Option Indicator. Enter ADD, AMEND, or CANCEL)
LINE 4 – ORDER ________________	(Order of Survey Accuracy)
LINE 5 – CONTROL POINT ________________	(Survey Control Point Name)
LINE 6 – UTM EAST ________________	(UTM .01-Meter Higher Order Easting)
LINE 7 – UTM NORTH ________________	(UTM .01-Meter Higher Order Northing)
LINE 8 – ALTITUDE ________________	(SCP Altitude to 0.1-Meter)
LINE 9 – SOURCE ________________	(Source of Station Data)
LINE 10 – STATION DESCRIPTION ________________	(Description, Control Point Station)
LINE 11 – ROUTE ________________	(Description of Route to SCP)
LINE 12 – ALFA ________________	(First Azimuth Mark Name)
LINE 13 – MARK ALFA ________________	(Description, Azimuth Mark)
LINE 14 – AZIMUTH TO ALFA ________________	(Azimuth to the Nearest One Thousandth of a Mil)
LINE 15 – BRAVO ________________	(Second Azimuth Mark Name)
LINE 16 – MARK BRAVO ________________	(Description, Azimuth Mark)
LINE 17 – AZIMUTH TO BRAVO ________________	(Azimuth to the Nearest One Thousandth of a Mil)
LINE 18 – NARRATIVE ________________	(Free Text for Additional Information Required for Clarification of Report)
LINE 19 – AUTHENTICATION ________________	(Report Authentication)

TITLE: SURVIVABILITY REPORT [SURREP]
REPORT NUMBER: S108

GENERAL INSTRUCTIONS: Use to report the completion of survivability tasks. Survivability tasks include digging, revetment, camouflage, and hardening of buildings. Reference: FM 5-34, FM 5-100, and FM 5-102.

LINE 1 – DATE AND TIME ________	(DTG)
LINE 2 – UNIT ________	(Unit Making Report)
LINE 3 – UNIT TYPE ________	(Type of Unit)
LINE 4 – TASK SERIAL NUMBER ________	(Unique Task Identifying Number)
LINE 5 – OBSTACLE ZONE/BELT NUMBER ________	(Zone, Belt Letter and Number)
LINE 6 – SURVIVABILITY MISSION ________	(Type of Mission, Protective Works, Camouflage)
LINE 7 – SURVIVABILITY TASK ________	(Task, CP, Fighting Positions)
LINE 8 – TYPE OF POSITION ________	(Type of Position)
LINE 9 – ECHELON ________	(Echelon of Unit)
LINE 10 – MAP SHEET NUMBERS ________	(Map Sheet Numbers Relevant to Report)
LINE 11 – NAME OF NEAREST TOWN ________	(Name of Nearby Town)
LINE 12 – COMPLETION TIME ________	(DTG When Task Was Completed)
LINE 13 – NARRATIVE ________	(Free Text for Additional Information Required for Clarification of Report)
LINE 14 – AUTHENTICATION ________	(Report Authentication)

TITLE: SYSTEM-REPLY OR REMARKS MESSAGE [SYS.RRM]
REPORT NUMBER: S110 {USMTF # F260}

GENERAL INSTRUCTIONS: Use to reply to a message received when a structured message does not exist or when it is necessary to provide plain text information. Reference: FM 11-32.

LINE 1 – DATE AND TIME ______________	(DTG)
LINE 2 – UNIT ______________	(Unit Making Report)
LINE 3 – REPORT TYPE ______________	(Report Type)
LINE 4 – ORIGINATOR ______________	(Originator)
LINE 5 – SERIAL ______________	(Report Serial Number)
LINE 6 – REPLY ______________	(Type of Reply Report. Enter ACKNOWLEDGE, WILL COMPLY, MODIFY EXECUTION, EXECUTION IMPOSSIBLE, VERIFICATION REQUESTED, EXECUTION COMPLETED)
LINE 7 – NARRATIVE ______________	(Free Text for Additional Information Required for Clarification of Report)
LINE 8 – AUTHENTICATION ______________	(Report Authentication)

TITLE: SYSTEM-REQUEST FOR REPORT [SYS.RFR]
REPORT NUMBER: S115 {USMTF # D260}

GENERAL INSTRUCTIONS: Use in fire support operations (1) to request a one-time fire support report or (2) to establish a standing requirement for a report. Reference: FM 6-71.

LINE 1 – DATE AND TIME ________________________________ (DTG)

LINE 2 – UNIT ________________________________ (Unit Making Report)

LINE 3 – POI ________________________________ (Primary Option Indicator. Enter ADD, AMEND, or CANCEL)

LINE 4 – PLAN ________________________________ (Fire Plan Designation)

LINE 5 – REPORT TYPE ________________________________ (Type of Fire Support Report)

LINE 6 – FREQUENCY ________________________________ (Report Frequency)

LINE 7 – FROM ________________________________ (DTG Effective From)

LINE 8 – TO ________________________________ (DTG Effective to)

LINE 9 – TARGET LIST ________________________________ (Target List/Last Target Indicator)

LINE 10 – GEOMETRY TYPE ________________________________ (Battlefield Geometry Type)

LINE 11 – GEOMETRY NAME ________________________________ (Battlefield Geometry Type Name)

LINE 12 – ELEMENT ________________________________ (Section, Platoon, Battery/Company, Battalion/Regiment, and Regiment/Brigade/Division Designators, or Ship Call Sign(s))

LINE 13 – NARRATIVE ________________________________ (Free Text for Additional Information Required for Clarification of Report)

LINE 14 – AUTHENTICATION ________________________________ (Report Authentication)

TITLE: TACTICAL ELINT REPORT [TACELINT]
REPORT NUMBER: T001 {USMTF # C121}

GENERAL INSTRUCTIONS: Use to report on the unit's tactical electronic intelligence systems. Reference: FM 34-10 and FM 34-10-13.

LINE 1 – DATE AND TIME ______________________________ (DTG)

LINE 2 – UNIT ______________________________ (Unit Making Report)

LINE 3 – TARGET ______________________________ (Target Identifier–BE Number, Target Signal Number, PIN Number, or RACKET Number)

LINE 4 – BETWEEN ______________________________ (DTG Detected)

LINE 5 – AND ______________________________ (DTG LOST or PRESENT if Still Emitting)

LINE 6 – EMITTER ______________________________ (Emitter Call Sign and Name or Nomenclature)

LINE 7 – LOCATION ______________________________ (UTM or Six-Digit Grid Coordinate With MGRS Grid Zone Designator)

LINE 8 – LENGTH ______________________________ (Length of Ellipse Semi-Major Axis)

LINE 9 – WIDTH ______________________________ (Width of Ellipse Semi-Minor Axis)

LINE 10 – AXIS ______________________________ (Axis Orientation)

LINE 11 – PARAMETERS FOLLOW ______________________________ (Alerts Addressee That the Following 10 Lines Contain Systems Parameters. If you are sending more than one set of parameters, state the number of the set being sent.)

EXAMPLE: PARAMETERS FOLLOW; SYSTEM ONE.

LINE 12 – FREQUENCY ______________________________ (Frequency of Signal)

LINE 13 – MODE ______________________________ (RF Operational Mode)

**Continued on next page.

TITLE: TACTICAL ELINT REPORT [TACELINT] (CONTINUED)
REPORT NUMBER: T001 {USMTF # C121}

LINE 14 – INTERVAL__________ (Pulse Repetition Interval)

LINE 15 – ACTIVITY __________ (Pulse Repetition Interval Activity)

LINE 16 – DURATION __________ (Pulse Duration)

LINE 17 – SCAN __________ (Scan Type)

LINE 18 – RATE __________ (Scan Type)

LINE 19 – POLARIZATION __________ (Antenna Polarization)

LINE 20 – BEARING__________ (Bearing of the Signal From the Detecting Unit's Position)

** Repeat lines 11 through 20 to report multiple mission/mission data. Assign sequential line numbers to succeeding iterations; for example, first iteration 11-15; second iteration 11a-15a; third iteration 11b-15b; and so on.

LINE 21 – NARRATIVE __________ (Free Text for Additional Information Required for Clarification of Report)

LINE 22 – AUTHENTICATION __________ (Report Authentication)

TITLE: TEMPORARY BURIAL SITE REQUEST [TEMPBURIALSITEREQ]
REPORT NUMBER: T005 {USMTF # B964}

GENERAL INSTRUCTIONS: Use to provide data required to request establishment of a temporary cemetery and special burial procedures/ceremonies. Reference: FM 10-296.

LINE 1 – DATE AND TIME ______________________	(DTG)
LINE 2 – UNIT ______________________	(Unit Making Report)
LINE 3 – REQUESTING AGENCY ______________________	(Identification of the Component/ Unit Making the Request)
LINE 4 – COORDINATES______________________	(UTM or Six-Digit Grid Coordinate With MGRS Grid Zone Designator of the Requesting Unit)
LINE 5 – SITE JUSTIFICATION ______________________	(Reason a Temporary Burial Site is Necessary)
LINE 6 – FEATURE______________________	(Prominent Terrain Feature of the Proposed Site)
LINE 7 – COORDINATES______________________	(UTM or Six-Digit Grid Coordinate With MGRS Grid Zone Designator of the Area Containing the Features Listed in the Preceding Line)
LINE 8 – NAME______________________	(Cemetery Name or Title)
LINE 9 – GRADIENT ______________________	(Relative Terrain Gradient (GENTLY ROLLING, MODERATE, STEEP, or VERY STEEP)
LINE 10 – DRAINAGE ______________________	(Type of Drainage (NATURAL, MAN-MADE, or NONE)
LINE 11 – SOIL ______________________	(Type of Soil)
LINE 12 – LAND USE______________________	(Current Use of the Land)
LINE 13 – ROAD NUMBER ______________________	(Access Road Number)

**Continued on next page.

TITLE: TEMPORARY BURIAL SITE REQUEST [TEMPBURIALSITEREQ] (CONTINUED)
REPORT NUMBER: T005 {USMTF # B964}

LINE 14 – ROAD NAME______________________________ (Name of Access Road to Cemetery Site)

LINE 15 – ROAD SURFACE ___________________________ (Type of Surface on Access Road)

LINE 16 – VEGETATION ______________________________ (Type of Vegetation)

LINE 17 – LOCAL LABOR______________________________ (Number of Local Laborers Available)

LINE 18 – EQUIPMENT ______________________________ (Type, Name, or Nomenclature of Available Cemetery Equipment)

LINE 19 – BURIAL TYPE______________________________ (Type of Special Burial Procedure/Ceremony Requested)

LINE 20 – PERSONNEL QUANTITY AND CLASS_____________ (Quantity and Classification of Remains (Military and Whether US, Allied, or Enemy))

**Transmit lines 19 and 20 for a special burial request.

**Repeat lines 19 through 20 to request multiple types of special burial procedures/ceremonies. Assign sequential line numbers to succeeding iterations; for example, first iteration 19 through 20; second iteration 19a through 20a; third iteration 19b through 20b; and so on.

LINE 21 – BURIAL JUSTIFICATION ______________________ (Reason for Requesting Mass Burial)

LINE 22 – NARRATIVE ______________________________ (Free Text for Additional Information Required for Clarification of Report)

LINE 23 – AUTHENTICATION __________________________ (Report Authentication)

TITLE: TRACK MANAGEMENT MESSAGE [TRKMAN]
REPORT NUMBER: T010 {USMTF # F752}

GENERAL INSTRUCTIONS: Use to report and resolve dual designations and differences in identity, category, and Identification Friend or Foe (IFF)/Selective Identification Feature (SIF). Reference: FM 44-100.

LINE 1 – DATE AND TIME ______________________________ (DTG)

LINE 2 – UNIT ______________________________ (Unit Making Report)

LINE 3 – CONFLICT ______________________________ (Type of Conflict: ID, CATEGORY, IFF, DUAL DESIGNATION, POSITION, ALTITUDE, SIZE, AIR OR SURFACE TYPE)

LINE 4 – TRACK ______________________________ (Track Number)

LINE 5 – POSITION______________________________ (UTM or Six-Digit Grid Coordinate With MGRS Grid Zone Designator)

LINE 6 – YOUR ______________________________ (Track Information)

LINE 7 – MY ______________________________ (Track Information)

LINE 8 – RESOLUTION ______________________________ (Resolution of Conflict)

LINE 9 – NARRATIVE ______________________________ (Free Text for Additional Information Required for Clarification of Report)

LINE 10 – AUTHENTICATION ______________________________ (Report Authentication)

TITLE: TRACK/POINT REPORT [TRKREP]
REPORT NUMBER: T015 {USMTF # F753}

GENERAL INSTRUCTIONS: Use to report, update, or drop air, surface, subsurface, and ground tracks or points in the absence of a TADIL A or TADIL B link. Reference: FM 44-100.

LINE 1 – DATE AND TIME ______________________	(DTG)
LINE 2 – UNIT ______________________	(Unit Making Report)
LINE 3 – ID ______________________	(Identification: FRIENDLY, UNKNOWN, HOSTILE)
LINE 4 – POSITION ______________________	(UTM or Six-Digit Grid Coordinate With MGRS Grid Zone Designator)
LINE 5 – TRACK ______________________	(Track Number)
LINE 6 – HEADING ______________________	(Direction of Travel in Degrees True or Magnetic)
LINE 7 – SPEED ______________________	(Speed of Movement in Knots)
LINE 8 – ALTITUDE ______________________	(Altitude in Hundreds of Feet)
LINE 9 – ENGAGED ______________________	(Engagement Status: YES or NO)
LINE 10 – NO. TYPE ______________________	(Number and Type of Craft: Two MIG 21s, Three Destroyers)
LINE 11 – SQUAWK ______________________	(IFF/SIF Mode and Code)
LINE 12 – MODE FOUR ______________________	(Confirmed Friend, No Response, Not Interrogated)
LINE 13 – CATEGORY ______________________	(Track Type: AIR, LAND, SURFACE, SUBSURFACE, or POINT)
LINE 14 – NARRATIVE ______________________	(Free Text for Additional Information Required for Clarification of Report)
LINE 15 – AUTHENTICATION ______________________	(Report Authentication)

TITLE: TRANSPORTATION SUPPORT REQUEST [TRANSSPTREQ]
REPORT NUMBER: T020 {USMTF # D825}

GENERAL INSTRUCTIONS: Use to request transportation support other than airlift support. Reference: FM 701-58.

LINE 1 – DATE AND TIME ______________________ (DTG)

LINE 2 – UNIT ______________________ (Unit Making Report)

LINE 3 – UNIT IDENTIFIER ______________________ (Identifier or Unit/Agency Requesting Transportation Support)

LINE 4 – CALL SIGN ______________________ (Call Sign of Unit)

LINE 5 – TELEX ______________________ (Unit/Agency Telex Number)

LINE 6 – UNIT POC ______________________ (Designated Unit POC)

LINE 7 – RANK/RATE ______________________ (POC Rank or Rate and Rating)

LINE 8 – PRIMARY PHONE ______________________ (Unit POC Primary Telephone Number)

LINE 9 – PRIMARY FREQUENCY ______________________ (Unit Primary Radio Frequency)

LINE 10 – POC LOCATION ______________________ (Unit POC Location Using UTM or Six-Digit Grid Coordinate With MGRS Grid Zone Designator)

LINE 11 – SECONDARY PHONE ______________________ (Unit POC Secondary Telephone Number)

LINE 12 – SECONDARY FREQUENCY ______________________ (Unit Secondary Radio Frequency)

LINE 13 – BILLING______________________ (Required Billing Instructions)

LINE 14 – PRIORITY ______________________ (Requestor-Assigned Priority Number)

LINE 15 – MODE______________________ (Requestor's Preferred Mode of Transportation)

**Continued on next page.

TITLE: TRANSPORTATION SUPPORT REQUEST [TRANSSPTREQ] (CONTINUED)
REPORT NUMBER: T020 {USMTF # D825}

LINE 16 – PICKUP TIME ______________________	(Desired Pickup DTG for Cargo/Passengers Being Moved)
LINE 17 – ON-LOAD ______________________	(UTM or Six-Digit Grid Coordinate With MGRS Grid Zone Designator of On-Load/Pickup Point)
LINE 18 – DELIVERY ______________________	(Desired Delivery Time for Cargo/Passengers Being Moved)
LINE 19 – OFF-LOAD ______________________	(UTM or Six-Digit Grid Coordinate With MGRS Grid Zone Designator of Off-Load Point)
LINE 20 – COUNT ______________________	(Count of Cargo/Passengers to be Transported)
LINE 21 – TYPE ______________________	(Type of Cargo/Passengers to be Transported)
LINE 22 – LENGTH ______________________	(Linear, End-to-End Measurement, of Each Type of Equipment)
LINE 23 – WIDTH ______________________	(Linear, Side-to-Side Measurement, of Each Type of Equipment)
LINE 24 – HEIGHT ______________________	(Vertical Dimension of Each Type of Equipment)
LINE 25 – WEIGHT ______________________	(Weight of Each Type of Equipment)
LINE 26 – HAZARDOUS DATA ______________________	(Required Hazardous Cargo Data According to Federal Regulation 49 And Joint Service Regulations)
LINE 27 – ADDITIONAL SUPPORT ______________________	(Additional Support Requirements at On-Load or Off-Load Points, if Applicable)

**Continued on next page.

TITLE: TRANSPORTATION SUPPORT REQUEST [TRANSSPTREQ] (CONTINUED)
REPORT NUMBER: T020 {USMTF # D825}

LINE 28 – POC FUNCTION ______________________________ (Name of the Function of the Mission POC)

LINE 29 – CALL SIGN ______________________________ (Call Sign of the Unit to Contact)

LINE 30 – REQUESTING POC______________________________ (Name of the Requesting Unit POC)

LINE 31 – RANK/RATE ______________________________ (Rank or Rate and Rating of Requesting POC)

LINE 32 – PRIMARY PHONE ______________________________ (Unit POC Primary Telephone Number)

LINE 33 – PRIMARY FREQUENCY ______________________________ (Primary Radio Frequency of Requesting Unit)

LINE 34 – POC LOCATION ______________________________ (UTM or Six-Digit Grid Coordinate With MGRS Grid Zone Designator of Reporting Unit POC)

LINE 35 – SECONDARY PHONE ______________________________ (Requesting POC Secondary Telephone Number)

LINE 36 – SECONDARY FREQUENCY ______________________________ (Secondary Radio Frequency of the Requesting Unit)

**Repeat lines 14 through 36 to initiate multiple requests for transportation support. Assign sequential line numbers to succeeding iterations; for example, first iteration 14 through 36; second iteration 14a through 36a; third iteration 14b through 36b; and so on.

LINE 37 – NARRATIVE ______________________________ (Free Text for Additional Information Required for Clarification of Report)

LINE 38 – AUTHENTICATION ______________________________ (Report Authentication)

TITLE: TRANSPORTATION SUPPORT RESPONSE [TRANSSPTRES]
REPORT NUMBER: T025 {USMTF # D826}

GENERAL INSTRUCTIONS: Use to reply to a transportation support request. Reference: FM 701-58.

LINE 1 – DATE AND TIME ______________________	(DTG)
LINE 2 – UNIT ______________________	(Unit Making Report)
LINE 3 – UNIT IDENTIFIER ______________________	(Identifier of Unit/Agency Responding for Transportation Support)
LINE 4 – CALL SIGN ______________________	(Call Sign of the Unit)
LINE 5 – TELEX ______________________	(Unit/Agency Telex Number)
LINE 6 – UNIT POC ______________________	(Designated Responding POC)
LINE 7 – RANK/RATE ______________________	(POC Rank or Rate and Rating)
LINE 8 – PRIMARY PHONE ______________________	(Unit POC Primary Telephone Number)
LINE 9 – PRIMARY FREQUENCY ______________________	(Responding Unit Primary Radio Frequency)
LINE 10 – POC LOCATION ______________________	(UTM or Six-Digit Grid Coordinate With MGRS Grid Zone Designator of Reporting Unit POC)
LINE 11 – SECONDARY PHONE ______________________	(Unit POC Secondary Telephone Number)
LINE 12 – SECONDARY FREQUENCY ______________________	(Responding Unit Secondary Radio Frequency)
LINE 13 – BILLING ______________________	(Required Billing Instructions)
LINE 14 – STATUS ______________________	(Disposition/Status Using APPROVED, DISAPPROVED, or MODIFIED, as Applicable)
LINE 15 – SUPPORT UNIT ______________________	(Supporting Unit Identifier or Designator, if Applicable)

**Continued on next page.

TITLE: TRANSPORTATION SUPPORT RESPONSE [TRANSSPTRES] (CONTINUED)
REPORT NUMBER: T025 {USMTF # D826}

LINE 16 – CALL SIGN ______________________	(Call Sign of the Supporting Unit)
LINE 17 – ON-LOAD SUPPORT EQUIPMENT NAME ___________	(Literal Name or Nomenclature of Support Equipment to be Provided at On-Load Point)
LINE 18 – ON-LOAD SUPPORT EQUIPMENT COUNT __________	(Number of Pieces of Support Equipment Assigned)
LINE 19 – OFF-LOAD SUPPORT EQUIPMENT NAME __________	(Literal Name or Nomenclature of Support Equipment to be Provided at Off-Load Point)
LINE 20 – OFF-LOAD SUPPORT EQUIPMENT COUNT _________	(Number of Pieces of Support Equipment Assigned)
LINE 21 – SUPPORT POC ______________________	(Supporting Unit POC Name)
LINE 22 – RANK/RATE ______________________	(Rank or Rate and Rating of Supporting POC)
LINE 23 – PRIMARY PHONE ______________________	(Supporting Unit POC Primary Telephone Number)
LINE 24 – PRIMARY FREQUENCY ______________________	(Primary Radio Frequency of the Supporting Unit)
LINE 25 – POC LOCATION ______________________	(UTM or Six-Digit Grid Coordinate With MGRS Grid Zone Designator of Reporting Unit POC)
LINE 26 – SECONDARY PHONE ______________________	(Supporting Unit POC Secondary Telephone Number)
LINE 27 – SECONDARY FREQUENCY ______________________	(Secondary Radio Frequency of the Supporting Unit)
LINE 28 – TRANSPORTATION MODE ______________________	(Mode of Transportation to be Provided, if the Mode is Other Than That Requested)

**Continued on next page.

TITLE: TRANSPORTATION SUPPORT RESPONSE [TRANSSPTRES] (CONTINUED)
REPORT NUMBER: T025 {USMTF # D826}

LINE 29 – PICKUP TIME ______________________________ (Pickup DTG for Movement, if the DTG is Other Than That Requested)

LINE 30 – ON-LOAD POINT ______________________________ (On-Load Point Location, if Other Than That Requested, Using UTM or Six-Digit Grid Coordinate With MGRS Grid Zone Designator)

LINE 31 – DELIVERY TIME ______________________________ (Delivery DTG for Movement, if the DTG is Other Than That Requested)

LINE 32 – OFF-LOAD POINT ______________________________ (Off-Load Point Location, if Other Than That Requested, Using UTM or Six-Digit Grid Coordinate With MGRS Grid Zone Designator)

LINE 33 – PRIORITY ______________________________ (Priority Assigned by Supporting Unit if it is Other Than That Requested)

**Repeat lines 14 to 33 to reply to different/multiple requests for transportation support. Assign sequential line numbers to succeeding iterations; for example, first iteration 14 through 33; second iteration 14a through 33a; third iteration 14b through 33b; and so on.

LINE 34 – NARRATIVE ______________________________ (Free Text for Additional Information Required for Clarification of Report)

LINE 35 – AUTHENTICATION ______________________________ (Report Authentication)

TITLE: UNIT SITUATION REPORT [UNITSITREP]
REPORT NUMBER: U001 {USMTF # C403}

GENERAL INSTRUCTIONS: Use to inform commanders of operational plans, unit readiness, and operational situations or summaries. Reference: FM 100-15, FM 71-100, FM 71-3, and FM 71-2.

LINE 1 – DATE AND TIME ______	(DTG)
LINE 2 – UNIT ______	(Unit Making Report)
LINE 3 – FROM ______	(Earliest DTG Zulu Report is Valid)
LINE 4 – UNTIL ______	(Latest DTG Zulu Report is Valid)
LINE 5 – MAP ______	(UTM or Six-Digit Grid Coordinate With MGRS Grid Zone Designator)
LINE 6 – ENEMY ______	(Enemy Activity Being Reported to Include: Nationality, Location, Mission, and Time of Sighting)
LINE 7 – NONHOSTILE ______	(Nonhostile Information Being Reported)
LINE 8 – OWN ______	(Activities of Own Forces Including Changes in Location of Units/Formations, and Activities of Forces Not Attached to Originating Unit)
LINE 9 – TYPE ______	(Type of Boundary Area or Line Being Described)
LINE 10 – POINT A ______	(Start Point for Boundary Line or Trace Being Described in UTM or Six-Digit Grid Coordinate with MGRS Grid Zone Designator)
LINE 11 – POINT B ______	(Second Point to Describe Line or Trace)
LINE 12 – POINT C ______	(Third Point to Describe Line or Trace)

**Continued on next page.

TITLE: UNIT SITUATION REPORT [UNITSITREP] (CONTINUED)
REPORT NUMBER: U001 {USMTF # C403}

LINE 13 – POINT D ______________________	(Fourth Point to Describe Line or Trace)
LINE 14 – POINT E ______________________	(Fifth Point to Describe Line or Trace)
LINE 15 – UNIT ______________________	(Unit Designator for the Reported Unit)
LINE 16 – LOCATION ______________________	(UTM or Six-Digit Grid Coordinate With MGRS Grid Zone Designator of Unit)
LINE 17 – UNITS ______________________	(Total Number of Units Being Reported)
LINE 18 – READY ______________________	(Total Number of Units Reporting C1, C2, or C3 Readiness)
LINE 19 – UNIT ______________________	(Unit Designator for Units With Less Than C3 Readiness Status)
LINE 20 – LOCATION ______________________	(UTM or Six-Digit Grid Coordinate With MGRS Grid Zone Designator of Unit)
LINE 21 – STATUS ______________________	(Degree to Which Unit is Combat Ready: C4 or C5)
LINE 22 – REASON ______________________	(Reason Unit has Less Than C3 Rating: RESOURCE, SUPPLY, TRAINING)
LINE 23 – READY ______________________	(Estimated Time Unit Will Achieve Combat Readiness Status)
LINE 24 – ADMINISTRATION ______________________	(Administration and Logistics Information Being Reported, to Include: KIA, WIA, MIA, NCW, and Equipment Lost/Damaged)

**Continued on next page.

TITLE: UNIT SITUATION REPORT [UNITSITREP] (CONTINUED)
REPORT NUMBER: U001 {USMTF # C403}

LINE 25 – GENERAL ____________________ (General Information Being Reported)

LINE 26 – SAFETY ____________________ (Commander's Safety Risk Assessment for Next Operation)

LINE 27 – NARRATIVE ____________________ (Free Text for Additional Information Required for Clarification of Report)

LINE 28 – AUTHENTICATION ____________________ (Report Authentication)

TITLE: US MEDICAL STATUS FIELD REPORT [USMEDFLDREP]
REPORT NUMBER: U005

GENERAL INSTRUCTIONS: Use to provide status of significant medical interests affecting the overall medical capability or health of command. Reference: FM 8-10-3.

LINE 1 – DATE AND TIME ______________________________ (DTG)

LINE 2 – UNIT ______________________________ (Unit Making Report)

LINE 3 – LOCATION ______________________________ (Reporting Unit UTM or Six-Digit Grid Coordinate With Grid Zone Designator)

LINE 4 – SHIP ______________________________ (Ship Name (if Required))

LINE 5 – BED STAT

a. OPERATING ______________________________ (Number Beds Operating)
b. NO. OCCUPIED ______________________________ (Number Beds Occupied)
c. NO. ARMY ______________________________ (Number Army Patients)
d. NO. NAVY ______________________________ (Number Navy Patients)
e. NO. MARINE ______________________________ (Number Marine Patients)
f. NO. AIR FORCE ______________________________ (Number Air Force Patients)
g. NO. EPW ______________________________ (Number EPW Patients)
h. NO. US CIVILIAN ______________________________ (Number US Civilian Patients)
i. NO. ALLIED CIVILIAN ______________________________ (Number Allied Civilian Patients)
j. NO. OTHER ______________________________ (Number Other Patients)

LINE 6 – STATUS ______________________________ (Patients Status by Service)

a. NO. RTD ______________________________ (Number RTD in 72 Hours)
b. NO. EVAC ______________________________ (Number EVAC to CONUS)
c. NO. DOW ______________________________ (Number DOW)

LINE 7 – DAYS CLASS VIII ______________________________ (Days of Class VIII OH)

LINE 8 – DAYS BLOOD ______________________________ (Days Blood Supply OH)

LINE 9 – CHRONIC PROBLEMS ______________________________ (Chronic Problems)

LINE 10 – NARRATIVE ______________________________ (Free Text for Additional Information Required for Clarification of Report)

LINE 11 – AUTHENTICATION ______________________________ (Report Authentication)

TITLE: WAR CRIME REPORTABLE INCIDENT REPORT [WCRIR]
REPORT NUMBER: W001

GENERAL INSTRUCTIONS: Use to convey flash traffic to the commander relating to a possible, suspected, or alleged violation of the law of war discovered within the command.

LINE 1 – DATE AND TIME ________________ (DTG)

LINE 2 – UNIT ________________ (Unit Making Report)

LINE 3 – TIME OF INCIDENT ________________ (Time the Incident Occurred)

LINE 4 – TIME OF DISCOVERY ________________ (Time the Incident Discovered)

LINE 5 – LOCATION OF INCIDENT ________________ (Where Incident Occurred/UTM or Six-Digit Grid Coordinates with MGRS Grid Zone Designator)

LINE 6 – PERSON(S) DISCOVERING ________________ (Names(s)/Unit(s) of Person(s) Discovering the Incident)

LINE 7 – SUMMARY ________________ (Narrative Description Including the Who, What, When, Where and How)

LINE 8 – UNITS IN AREA ________________ (Identity of Location and Point of Contact for All Evidence)

LINE 9 – LOCATION OF EVIDENCE ________________ (Identity of Location and Point of Contact for All Evidence)

LINE 10 – AUTHENTICATION ________________ (Report Authentication)

TITLE: WARNING MESSAGE-AIR DEFENSE [AIRDEFWARN]
REPORT NUMBER: W003 {USMTF # F715}

GENERAL INSTRUCTIONS: Use to transmit air defense warnings and weapons control conditions. Reference: FM 44-100.

LINE 1 – DATE AND TIME ____________________	(DTG)
LINE 2 – UNIT ____________________	(Unit Making Report)
LINE 3 – WARNING ____________________	(Air Defense Warning: WHITE, YELLOW, or RED)
LINE 4 – STATUS ____________________	(Weapon Control Status: Free, Tight, or Hold)
LINE 5 – EFFECTIVE ____________________	(Effective DTG; Transmit Only if Effective Time is Not Immediate)
LINE 6 – AREA ____________________	(Area or Sector Affected)
LINE 7 – BY ____________________	(Call Sign of Declaring Authority)
LINE 8 – NARRATIVE ____________________	(Free Text for Additional Information Required for Clarification of Report)
LINE 9 – AUTHENTICATION ____________________	(Report Authentication)

TITLE: WARNING ORDER [WARNORD]
REPORT NUMBER: W005

GENERAL INSTRUCTIONS: Use to convey a preliminary notice of an impending order or action. Reference: FM 101-5-1 and FM 101-5.

LINE 1 – DATE AND TIME ______________________ (DTG)

LINE 2 – UNIT ______________________ (Unit Making Report)

LINE 3 – WARNING ORDER ______________________ (Warning Order Number)

LINE 4 – REFERENCES ______________________ (References)

LINE 5 – TIME ZONE ______________________ (TZ Used in WARNORD)

LINE 6 – TASK ORGANIZATION ______________________ (Optional)

LINE 7 – SITUATION

a. ENEMY ______________________ (Enemy Forces)

b. FRIENDLY ______________________ (Friendly Forces)

c. ATTACH/DET ______________________ (Attach/Detachments)

LINE 8 – MISSION ______________________

LINE 9 – EXECUTION INTENT ______________________

a. CONCEPT ______________________ (Concept of Operation)

b. MAN TASKS ______________________ (Tasks to Maneuver Units)

c. CS TASKS ______________________ (Tasks to CS Units)

d. INSTRUCTIONS ______________________ (Coordinating Instructions)

(1) CCIR ______________________

(2) Risk MGT ______________________

(3) Deception ______________________

(4) Priorities ______________________

(5) Time Line ______________________

(6) Rehearsals ______________________

(7) Orders GRP MTG ______________________

(8) Earliest MVMT TIME ______________________

**Continued on next page.

TITLE: WARNING ORDER [WARNORD] (CONTINUED)
REPORT NUMBER: W005

LINE 10 – SERVICE AND SUPPORT

a. SPECIAL EQUIPMENT____________________ (Special Equipment)

b. TRANSPORTATION____________________ (Transportation)

LINE 11 – COMMAND AND SIGNAL

a. COMMAND ____________________ (Command)

b. SIGNAL ____________________ (Signal)

LINE 12 – ACKNOWLEDGE ____________________ (Mandatory)

LINE 13 – CDR NAME/RANK ____________________ (Commander's Name and Rank)

LINE 14 – NARRATIVE ____________________ (Free Text for Additional Information Required for Clarification of Report)

LINE 15 – AUTHENTICATION ____________________ (Report Authentication)

TITLE: WATER SUPPLY POINT [WTRSUPPT]
REPORT NUMBER: W010 {USMTF # F864}

GENERAL INSTRUCTIONS: Use to provide operational information about the water supply. Reference: FM 10-52.

LINE 1 – DATE AND TIME ______________________	(DTG)
LINE 2 – UNIT ______________________	(Unit Making Report)
LINE 3 – SUPPORT UNIT ______________________	(Designation of the Supporting Unit)
LINE 4 – FROM ______________________	(DTG for Beginning of Period Applying to Operational Information)
LINE 5 – TO ______________________	(DTG for End of Period Applying to Operational Information)
LINE 6 – POTABLE LOCATION ______________________	(UTM or Six-Digit Grid Coordinate With MGRS Grid Zone Designator of Potable Water Supply Point)
LINE 7 – POTABLE STATUS ______________________	(OPERATIONAL, NOT OPERATIONAL, or LIMITED)
LINE 8 – POTABLE AVAILABLE ______________________	(Gallons of Potable Water Available at Water Supply Point)
LINE 9 – POINT CAPACITY ______________________	(Daily Production Capacity in Gallons)
LINE 10 – WATER UNIT ______________________	(Name or Designator of Unit Providing Support Shown in Line 11)
LINE 11 – NONPOTABLE LOCATION ______________________	(UTM or Six-Digit Grid Coordinate With MGRS Grid Zone Designator of Nonpotable Water Supply Point)
LINE 12 – NONPOTABLE AVAILABLE ______________________	(Gallons of Nonpotable Water Available at Water Supply Point)

**Continued on next page.

TITLE: WATER SUPPLY POINT [WTRSUPPT] (CONTINUED)
REPORT NUMBER: W010 {USMTF # F864}

**Repeat lines 4 through 13 if necessary. Assign sequential line numbers to succeeding iterations; for example, first iteration 4 through 13; second iteration 4a through 13a; third iteration 4b through 13b; and so on.

LINE 13 – NONPOTABLE CAPACITY ________________________ (Daily Production Capacity in Gallons)

LINE 14 – NARRATIVE ___________________________________ (Free Text for Additional Information Required for Clarification of Report)

LINE 15 – AUTHENTICATION _____________________________ (Report Authentication)

TITLE: WEATHER ADVISORY/WATCH [WEATHERWATCH]
REPORT NUMBER: W015

GENERAL INSTRUCTIONS: Use to send flash weather information that will affect current unit operations. Reference: FM 34-81.

LINE 1 – DATE AND TIME	________________	(DTG)
LINE 2 – UNIT	________________	(Unit Making Report)
LINE 3 – REPORT LINE	________________	(Report Line From 0117)
LINE 4 – SUMMARY	________________	(Summary of Warning)
LINE 5 – TIME OF WATCH	________________	(DTG From-to DTG as of DTG of Watch)
LINE 6 – AREA	________________	(Area Affected)
LINE 7 – NARRATIVE	________________	(Free Text for Additional Information Required for Clarification of Report)
LINE 8 – AUTHENTICATION	________________	(Report Authentication)

TITLE: WEATHER FORECAST [WXFCST]
REPORT NUMBER: W020 {USMTF # C521}

GENERAL INSTRUCTIONS: Use to provide the commander with forecasted weather expected in the area of operation until normal weather communications can be established. Update forecasts at least every six hours. Reference: FM 34-81.

LINE 1 – DATE AND TIME ______________________	(DTG)
LINE 2 – UNIT ______________________	(Unit Making Report)
LINE 3 – LOCATION ______________________	(Center of Forecasted Weather Expressed Using UTM or Six-Digit Grid Coordinate With MGRS Grid Zone Designator)
LINE 4 – VARIATION ______________________	(Weather Variation With Respect to Time: INTERMITTENT, GRADUAL, or TEMPORARY)
LINE 5 – VALID ______________________	(DTG Zulu Forecast is Effective)
LINE 6 – UNTIL ______________________	(DTG Zulu Forecast is No Longer in Effect)
LINE 7 – CEILING ______________________	(Lowest Forecast Ceiling in Hundreds of Feet Above Ground Level)
LINE 8 – COVER ______________________	(Total Sky Coverage in Eighths)
LINE 9 – VISBY ______________________	(Prevailing Visibility in Meters)
LINE 10 – WEATHER ______________________	(Type of Weather Phenomena Being Forecast)
LINE 11 – MAX ______________________	(Maximum Temperature Forecasted in Degrees Celsius)
LINE 12 – MIN ______________________	(Minimum Temperature Forecasted in Degrees Celsius)
LINE 13 – FREEZE LEVEL ______________________	(Minimum Freezing Level in Thousands of Feet)

**Continued on next page.

TITLE: WEATHER FORECAST [WXFCST] (CONTINUED)
REPORT NUMBER: W020 {USMTF # C521}

LINE 14 – WIND	____________	(Direction and Variable Wind Direction in Degrees)
LINE 15 – SPEED	____________	(Maximum Sustained Wind Speed in Knots)
LINE 16 – GUSTS	____________	(Peak Gusts in Knots)
LINE 17 – ALTIMETER	____________	(Altimeter Setting in Hundredths of Inches of Mercury)
LINE 18 – WIND@2000ft	____________	(Wind Direction and Speed at 2000 Feet)
LINE 19 – WIND@5000ft	____________	(Wind Direction and Speed at 5000 Feet)
LINE 20 – WIND@10,000ft	____________	(Wind Direction and Speed at 10,000 Feet)
LINE 21 – WIND@15,000ft	____________	(Wind Direction and Speed at 15,000 Feet)
LINE 22 – WIND@20,000ft	____________	(Wind Direction and Speed at 20,000 Feet)
LINE 23 – NARRATIVE	____________	(Free Text for Additional Information Required for Clarification of Report)
LINE 24 – AUTHENTICATION	____________	(Report Authentication)

SECTION III—CROSS-OVER INDEXES

Intelligence Reports and Messages

Maneuver Reports and Messages

Rules of Engagement Reports and Messages

Service Support Reports and Messages

Signal Reports and Messages

GLOSSARY

Below are acronyms and abbreviations of commonly used terms in FM 101-5-2.

A&L	administration/logistics
A2C2	Army airspace command and control
AA	avenue of approach
ABCS	Army Battle Command System
ACO	airspace control order
ADA	air defense artillery
AE	aeromedical education
AH	attack helicopter
AM	amplitude modulated
AO	area of operations
AR	Army regulation
auth	authorized
AWOL	absent without leave
BE	basic encyclopedia
bio	biological
BP	battle position
btl	battle
C2	command and co ntrol
CA	civil affairs; chaplain's assistant
CAS	close air support
CCIR	commander's critical information requirements
cdr	commander
cGy	centigray
CH	cargo helicopter
Ch	chaplain
CHEM	chemical
CHOP	change of operational control procedure
CINC	commander in chief
cl	class
CM	countermobility
cmd	command
CMO	civil-military operations
CO	company
COA	course of action
COMINT	communications intelligence
CONUS	continental United States
COOP	Continuity of Operations Plan
CP	command post
CPL	corporal
CS	combat support
CSM	command sergeant major
CSS	combat service support
DA	Department of the Army
decon	decontamination
dir	direction
DOS	days of supply
DOW	died of wounds
DS	direct support
DTG	date-time group
EA	engagement area; electronic activity
EAC	echelons above corps
EDF	effective direct fire
EDM	effective downwind message
EEI	essential elements of information
enl	enlisted
EOD	explosive ordinance disposal
EPW	enemy prisoner of war
ES	electronic surveillance; electronic warfare support
EVAC	evacuation
EW	electronic warfare; early warning
FA	field artillery
FASCAM	family of scatterable mines
FEBA	forward edge of the battle area
FLOT	forward line of own troops
FM	field manual; fire mission
FN	foreign nation
FRAGO	fragmentary order
G4	Assistant Chief of Staff, G4 (Logistics)
Gator	A scatterable mine system delivered by Air Force and Navy tactical aircraft
GEOLOC	geographic location
GEOREF	World Geographic Reference
GS	general support
GSR	general support reinforcing ;
gyn	gynecology

H-hour	specific time at which an operation or exercise is scheduled to start
HUMINT	human intelligence
IAW	in accordance with
ICAO	International Civil Aviation Organization
ICM	improved conventional munitions; intelligence collection management
ID	identification
IEW	intelligence and electronic warfare
IFF	identification friend or foe
IMINT	imagery intelligence
intel	intelligence
INTREP	intelligence report
IR	information requirement
J-3	Operations Directorate of joint staff
J-6	Command, Control, Communications, and Computer System Directorate of a joint staff
JMEM	Joint Munitions Effectiveness Manual
JOC	Joint Operations Center
JP	jet petroleum; joint publication
JRCC	joint rescue coordination center
KIA	killed in action
KM	kilometer
KT	kilotons
lat	latitude
LD	line of departure
LN.XX	liaison
LOGPAC	logistics package
long	longitude
LRP	Logistics Release Point
LZ	landing zone
MASINT	measurement and signature intelligence
MBBLs	thousands of barrels
MC	pediatric
MCM	mine countermeasures
MEDEVAC	medical evacuation
MEDREGREP	Medical Regulating Report
MEDSTAT	Medical Status Report
met	meteorological
MET.CF	Meteorological - Fallout Message
MET.CM	Meteorological - Computer Message
MET.TA	Meteorological - Target Acquisition Message
MFR	memorandum for record
MGT	management
MHz	megahertz
MIA	missing in action
MICLIC	mine-clearing line charge
MIJI	meaconing, intrusion, jamming and interference
MLRS	multiple launch rocket system
MOGAS	motor gasoline
MOPMS	Modular Pack Mine System
MOPP	mission-oriented protective posture
MOS	military occupational specialty
MP	psychiatric
MPH	miles per hour
MSD	minimum safe distance
msn	mission
MSR	main supply route
MT	megaton
MTE	medical treatment element
mvmt	movement
NATO	North Atlantic Treaty Organization
NAV	navigation
NBC	nuclear, biological, and chemical
NCA	National Command Authorities
NCO	noncommissioned officer
NCW	not complied with
NE	Northeast; Netherlands
NMC	not mission capable
no	number
NOH	number on hand
NSF	naval surface fire
NSN	national stock number
NUC	nuclear
NUDETSUM	nuclear detonations summary report
O&I	operations and intelligence
OB	obstetrics; order of battle

off	officer
OH	on hand
OPREP	operational report
OR	operational readiness
PA	public affairs
pam	pamphlet
PAO	public affairs office/officer
PID	plan identification number
PIR	priority intelligence requirements
PM	provost marshal
POC	Point of Contact
POI	primary option indictor
POL	petroleum, oils, and lubricants
PSG	platoon sergeant
PSS	personnel service support
PSYOP	psychological operations
PZ	pickup zone
qty	quantity
R	reinforcing
RAP	rear area protection
RCC	rescue coordination center
RDD	required delivery date
recon	reconnaissance
RFI	request for information
ROC	rear operations center
ROE	rules of engagement
RTD	return to duty
S4	logistics staff officer brigade and below
SAEDA	Subversion and Espionage Directed Against the Army
SAR	search and rescue
SB	burn
SC	spinal cord
SENREP	sensor report
SF	Special Forces
SG	obstetrics/gynecology
SIF	selective identification feature
SIGINT	signal intelligence
SO	orthopedic
SOP	standing operating procedures
SPIREP	Spot Intelligence Report
SRI	Standing Request for Information
SS	surgery
SSC	thoracic
SSM	maxillo
SSN	neuro; social security number
SSO	ophthal
sta	station
STRIKWARN	strike warning
tac	tactical
TACAN	Tactical Air Navigation
TADIL	Tactical Air Defense Information Link
TCN	transportation control number
temp	temperature
tgt	target
TM	target material
TOA	trade-off analysis; Transportation Operating Agency
TRADOC	US Army Training and Doctrine Command
UAV	unmanned aerial vehicle
UMT	unit ministry team
US	United States
USAF	United States Air Force
UTM	Universal Transverse M ercator
VIP	very important person
WARNORD	warning order
WIA	wounded in action
WILCO	will comply with transmission
WO	warrant officer

REFERENCES

Joint and Multinational Publications

ATP-45 (A) *Reporting Nuclear Detonations, Biological and Chemical Attacks, and Predicting and Warning of Associated Hazards and Hazard Areas.*

JP 3-56.1 *Command and Control for Joint Air Operations*, 14 November 1994.

STANAG 2434 *Allied Procedural Publication 9:A Compendium of Allied Land Forces Messages*, Edition 1, 1997 (FOUO).

Army Regulations

AR 190-40 *Serious Incident Report*, 30 November 1993.

AR 310-50 *Authorized Abbreviations, Brevity Codes, and Acronyms*, 15 November 1985.

AR 385-40 *Accident Reporting and Records*, 01 November 1994.

Field Manuals

FM 1-100 *Army Aviation Operations*, 21 February 1997.

FM 1-108 *Doctrine for Army Special Operations Aviation Forces*, 03 November 1993.

FM 1-114 *Tactics, Techniques, and Procedures for the Regimental Aviation Squadron*, 20 February 1991.

FM 1-513 *Battlefield Recovery and Evacuation of Aircraft*, 20 May 1993.

FM 3-3 *Chemical and Biological Contamination Avoidance*, 16 November 1992.

FM 3-3-1 *Nuclear Contamination Avoidance*, 09 September 1994.

FM 3-4 *NBC Protection*, 29 May 1992.

FM 3-5 *NBC Decontamination*, 17 November 1993.

FM 3-7 *NBC Field Handbook*, 29 September 1994.

FM 3-14 *Nuclear, Biological, and Chemical (NBC) Vulnerability Analysis*, 12 November 1997.

FM 5-34 *Engineer Field Data*, 14 September 1987.

FM 5-100 *Engineer Operations*, 27 February 1996.

FM 5-101 *Mobility*, 23 January 1985.

FM 5-102 *Countermobility*, 14 March 1985.

FM 6-2 *Tactics, Techniques, and Procedures for Field Artillery Survey*, 23 September 1993.

FM 6-16 *Tables for Artillery Meteorology (Electronic) Ballistic Type 3 and Computer Messages*, 10 May 1979.

FM 6-16-1 *Tables for Artillery Meteorology (Sound Ranging)* Messages, 10 May 1979.

FM 6-16-2 *Tables for Artillery Meteorology (Visual) Ballistic Type 3 and Computer Messages and Limited Surface Observations*, 15 January 1982.

FM 6-16-3 *Tables for Artillery Meteorology (Electronic and Visual) Type 2 Messages*, 15 June 1982.

FM 6-20 *Fire Support in the Airland Battle*, 17 May 1988.

FM 6-20-10 *Tactics, Techniques, and Procedures for the Targeting Process*, 08 May 1996.

FM 6-20-30 *Tactics, Techniques, and Procedures for Fire Support for Corps and Division Operations*, 18 October 1989.

FM 6-20-40 *Tactics, Techniques, and Procedures for Fire Support for Brigade Operations (Heavy)*, 05 January 1990.

FM 6-30 *Tactics, Techniques, and Procedures for Observed Fire*, 16 July 1991.

FM 6-71 *Tactics, Techniques, and Procedures for Fire Support for the Combined Arms Commander*, 29 September 1994.

FM 6-121 *Tactics, Techniques, and Procedures for Field Artillery Target Acquisition*, 25 September 1990.

FM 7-7 *The Mechanized Infantry Platoon and Squad*, 15 March 1985.

FM 7-92 *The Infantry Reconnaissance Platoon and Squad (Airborne, Air Assault, Light Infantry)*, 23 December 1992.

FM 8-10 *Health Service Support in a Theater of Operations*, 01 March 1991.

FM 8-10-1 *Tactics, Techniques, and Procedures for the Medical Company*, 29 December 1994.

FM 8-10-3 *Division Medical Operations Center Tactics, Techniques, and Procedures*, 12 November 1996.

FM 8-10-5 *Brigade and Division Surgeon's Handbook Tactics, Techniques, and Procedures*, 10 June 1991.

FM 8-10-6 *Medical Evacuation in a Theater of Operations, Tactics, Techniques, and Procedures*, 31 October 1991.

FM 8-10-9 *Combat Health Logistics in a Theater of Operations Tactics, Techniques, and Procedures*, 03 October 1995.

FM 8-10-24 *Area Support Medical Battalion Tactics, Techniques, and Procedures*, 13 October 1993.

FM 9-6 *Munitions Support in Theater of Operations*, 20 March 1998.

FM 9-15 *Explosive Ordnance Disposal Service and Unit Operations*, 08 May 1996.

FM 9-43-1 *Maintenance Operations and Procedures*, 28 September 1997.

FM 10-52 *Water Supply in Theaters of Operations*, 11 July 1990.

FM 10-63-1 *Graves Registration Handbook*, 17 July 1986.

FM 10-286 *Identification of Deceased Personnel*, 30 June 1976.

FM 10-296 *Headquarters and Headquarters Units, Graves Registration Battalion*, 10 February 1986.

FM 10-297 *Graves Registration Company*, 25 April 1986.

FM 11-1 *Multiservice Procedures for the Single Channel Ground and Airborne Radio System*, 29 May 1996.

FM 11-32 *Combat Net Radio Operations*, 15 October 1990.

FM 12-6 *Personnel Doctrine*, 09 September 1994.

FM 16-1 *Religious Support*, 26 May 1995.

FM 17-12-1/2 *Tank Gunnery (Abrams) Volumes I & II*, 19 March 1993.

FM 17-15 *Tank Platoon*, 03 April 1996.

FM 17-95 *Cavalry Operations*, 24 December 1996.

FM 17-98 *Scout Platoon*, 09 September 1994.

FM 19-1 *Military Police Support for the Airland Battle*, 23 May 1988.

FM 19-30 *Physical Security*, 01 March 1979.

FM 19-40 *Enemy Prisoners of War, Civilian Internees, and Detained Persons*, 27 February 1976.

FM 20-32 *Mine/Countermine Operations*, 30 September 1992.

FM 21-26 *Map Reading and Land Navigation*, 07 May 1993.

FM 23-1 *Bradley Gunnery*, 18 March 1996.

FM 33-1	*Psychological Operations*, 18 February 1993.
FM 34-1	*Intelligence and Electronic Warfare Operations*, 27 September 1994.
FM 34-2	*Collection Management and Synchronization Planning*, 08 March 1994.
FM 34-2-1	*Tactics, Techniques, and Procedures for Reconnaissance and Surveillance and Intelligence Support to Counterreconnaissance*, 19 June 1991.
FM 34-3	*Intelligence Analysis*, 15 March 1990.
FM 34-10	*Division Intelligence and Electronic Warfare Operations*, 25 November 1986.
FM 34-10-2	*Intelligence and Electronic Warfare (IEW) Equipment Handbook*, 13 July 1993.
FM 34-10-13	*Tactical Control and Analysis Center (TCAC) AN/TSQ-130 (V)*, 14 August 1989.
FM 34-25-3	*All-Source Analysis System and the Analysis and Control Element*, 03 October 1995.
FM 34-36	*Special Operations Forces Intelligence and Electronic Warfare*, 30 September 1991.
FM 34-40	*Electronic Warfare Operations*, 09 October 1987.
FM 34-37	*Echelons Above Corps Intelligence and Electronic Warfare Operations*, 15 January 1991.
FM 34-40-7	*Communications Jamming Handbook*, 23 November 1992.
FM 34-40-9	*Direction Finding Operations*, 29 August 1991.
FM 34-54	*Battlefield Technical Intelligence*, 05 April 1990.
FM 34-60	*Counterintelligence*, 03 October 1995.
FM 34-80	*Brigade and Battalion Intelligence and Electronic Warfare Operations*, 15 April 1986.
FM 34-81	*Weather Support for Army Tactical Operations*, 01 August 1989.
FM 41-10	*Civil Affairs Operations*, 11 January 1993.
FM 44-100	*US Army Air Defense Operations*, 15 June 1995.
FM 46-1	*Public Affairs Operations*, 30 May 1997.
FM 63-11	*Logistics Support Element Tactics, Techniques, and Procedures*, 08 October 1996.
FM 63-20	*Forward Support Battalion*, 26 February 1990.
FM 71-2	*The Tank and Mechanized Infantry Battalion Task Force*, 27 September 1988.
FM 71-3	*The Armored and Mechanized Infantry Brigade*, 08 January 1996.
FM 71-100	*Division Operations*, 28 August 1996.
FM 90-13-1	*Combined Arms Breaching Operations*, 28 February 1991.
FM 90-70	*Combined Arms Obstacle Integration*, 29 September 1994.
FM 100-15	*Corps Operations*, 29 October 1996.
FM 100-103	*Army Airspace Command and Control in a Combat Zone*, 07 October 1987.
FM 100-103-1	*ICAC2 Multiservice Procedures for Integrated Combat Airspace Command and Control*, 03 October 1994.
FM 100-103-2	*Multiservice Procedures for the Theater Air-Ground System (TAGS)*, 31 October 1994.
FM 101-5	*Staff Organization and Operations*, 31 May 1997.
FM 101-5-1	*Operational Terms and Graphics*, 30 September 1997.
FM 700-80	*Logistics*, 15 August 1985.
FM 701-58	*Planning Logistics Support for Military Operations*, 27 May 1987.

Department of the Army Pamphlets

DA PAM 25-7, US Message Text Formats, January 1998.

Made in the USA
Monee, IL
04 November 2026

67383146R10174